Global Risks and Crises Management in Tourism

Elbeyi Pelit / Hasan Hüseyin Soybali / Ali Avan (eds.)

Global Risks and Crises Management in Tourism

Theoretical and Practical Perspectives

Bibliographic Information published by the
Deutsche Nationalbibliothek
The Deutsche Nationalbibliothek lists this publication in the Deutsche
Nationalbibliografie; detailed bibliographic data is available online at
http://dnb.d-nb.de.

Library of Congress Cataloging-in-Publication Data
A CIP catalog record for this book has been applied for at the
Library of Congress.

Cover illustration: © Photo by Bruno Cervera from Pexels

This book has been published with the support of InGlobe Academy.
www.inglobe.org

ISBN 978-3-631-85842-4 (Print)
E-ISBN 978-3-631-87113-3 (E-PDF)
E-ISBN 978-3-631-87200-0 (EPUB)
10.3726/b19362

© Peter Lang GmbH
Internationaler Verlag der Wissenschaften
Berlin 2021
All rights reserved.

Peter Lang – Berlin · Bern · Bruxelles · Istanbul · New York · Oxford · Warszawa · Wien

All parts of this publication are protected by copyright. Any
utilisation outside the strict limits of the copyright law, without
the permission of the publisher, is forbidden and liable to
prosecution. This applies in particular to reproductions,
translations, microfilming, and storage and processing in
electronic retrieval systems.

This publication has been peer reviewed.

www.peterlang.com

Preface

Tourism is one of the most sensitive industries across risks and crises based on economic, social, and political developments. In recent years, the challenges such as terrorism, epidemics, and economic blockades based on political tension between countries, etc., cause increased risks and threats on destinations but also cause decreased touristic mobility. A global pandemic, depressed economy, political uncertainties, and social problems can also interrupt the touristic movement. For all these reasons, organizations in the tourism industry strive to build a trustworthy atmosphere by minimizing risks and protecting and improve destination image. Risk and crises management is of great importance in anticipating possible threats and dangers, eliminating uncertainties, taking precautions before problems occur, and managing the process correctly during emergencies. Managing risks, uncertainties, and crises in the tourism industry are essential compared to other sectors due to its fragile structure to cyclical developments, needing substitution, and spatiality. From this point of view, with this chapter focusing on global risks and crises management in the tourism industry with a proactive approach, it is aimed to create an essential source/reference about preparation to emergencies, steps to be taken to eliminate uncertainties and risks, and effective risk and crisis management practices, and to guide managers and industry practitioners.

Dr. Elbeyi PELİT, Dr. H. Hüseyin SOYBALI, Dr. Ali AVAN

Contents

List of Contributors 9

Ahmet BAYTOK, Ali AVAN and Özcan ZORLU
What We Have Learned From COVID-19? Managerial Advantages of Crisis in Tourism 13

Bayram ŞAHIN, Selda GÜVEN and İbrahim MISIR
Crisis, Terrorism, Epidemic Diseases and Their Impacts on International Destination Selection 27

Selcen Seda TURKSOY
Domestic Tourism Post Covid-19 – An Opportunity to Revive Turkish Tourism? 65

İbrahim Tolga ÇOŞKUN
Economic Effects of Tourism and Tourism in the Period of Economic Crises 83

Berna (KIRAN) BULĞURCU
Probing the Efficiency of Tourism Sector before and after Economic Crisis Periods: The Case of European Union Countries 99

Nesrin ÖZKAN
The Impact of COVID-19 on City Indices in Turkey: An Event Study Analysis 123

Alper ATEŞ and Ömür Hakan KUZU
Tourism Higher Education Management and Policies in Times of Crisis 147

List of Figures 165

List of Tables 167

Notes on Contributors 169

About Editors 173

List of Contributors

Editors
Prof. Dr. Elbeyi PELİT
https://orcid.org/0000-0002-6418-801X
Afyon Kocatepe University
Faculty of Tourism
elbeyipelit@aku.edu.tr
Prof. Dr. Hasan Hüseyin SOYBALI
https://orcid.org/0000-0002-5929-0933
Afyon Kocatepe University
Faculty of Tourism
hsoybali@aku.edu.tr
Assoc. Prof. Ali AVAN
https://orcid.org/0000-0003-4510-3962
Afyon Kocatepe University
Faculty of Tourism
aliavan@aku.edu.tr

Chapter 1
Assoc. Prof. Ahmet BAYTOK
https://orcid.org/0000-0002-5826-7694
Afyon Kocatepe University,
Faculty of Tourism
ahmetbaytok@aku.edu.tr
Assoc. Prof. Ali AVAN
https://orcid.org/0000-0003-4510-3962
Afyon Kocatepe University
Faculty of Tourism
aliavan@aku.edu.tr
Assoc. Prof. Özcan ZORLU
https://orcid.org/0000-0003-3533-1945
Afyon Kocatepe University
Faculty of Tourism
ozcanzorlu@aku.edu.tr

Chapter 2
Assoc. Prof. Bayram ŞAHIN,
https://orcid.org/0000-0002-1911-9066
Balıkesir University
Faculty of Tourism
bsahin@balikesir.edu.tr
Lecturer Selda GÜVEN
https://orcid.org/0000-0002-4931-7880
Çanakkale Onsekiz Mart University
Ezine Vocational School
seldaegilmezgil@comu.edu.tr
Res. Asst. İbrahim MISIR
https://orcid.org/0000-0002-8233-6346
Osmaniye Korkut Ata University
Kadirli School of Applied Sciences
ibrahimmisir@balikesir.edu.tr

Chapter 3
Asst. Prof. Selcen Seda TURKSOY
https://orcid.org/0000-0003-1842-0858
Ege University
Faculty of Tourism, Çeşme
selcen.seda.turksoy@ege.edu.tr

Chapter 4
Res. Asst. İbrahim Tolga ÇOŞKUN
https://orcid.org/0000-0002-5314-3748
Çukurova University
Faculty of Economics and Administrative Sciences
tcoskun@cu.edu.tr

Chapter 5
Assoc. Prof. Berna (KIRAN) BULĞURCU
https://orcid.org/0000-0002-9695-2668
Çukurova University
Faculty of Economics and Administrative Sciences
bkiran@cu.edu.tr

Chapter 6
Nesrin ÖZKAN
https://orcid.org/0000-0002-8674-5518
Atlas University
Faculty of Humanities and Social Sciences
nesrin.ozkan@atlas.edu.tr

Chapter 7
Assoc. Prof. Alper ATEŞ
https://orcid.org/0000-0002-4347-7306
Selçuk University
Faculty of Tourism
alpera@selcuk.edu.tr
Assoc. Prof. Ömür Hakan KUZU
https://orcid.org/0000-0002-2996-0721
Selçuk University
Beyşehir Ali Akkanat Faculty of Tourism
ohkuzu@selcuk.edu.tr

Ahmet BAYTOK, Ali AVAN and Özcan ZORLU

What We Have Learned From COVID-19? Managerial Advantages of Crisis in Tourism

1 Introduction

In tourism behavior, visitors travel with a variety of motivators (psychical, emotional, personal, cultural, personal development, and status) which trigger the economic aspect of the tourism industry (Swarbrooke & Horner, 2007: 54). However, these motivators cannot guarantee ever-growing international tourist arrivals and correspondingly stable growth of tourism supply. Hence, tourism demand requires not only the existence of disposable income but also appropriate conditions such as political stability, comfortable and safe transportation, proper destination management, and high-quality tourism services. Further, visitors in tourism desire to feel themselves in safety and healthy atmosphere. Thus, a chaotic atmosphere or comfortless conditions directly decrease the number of tourist arrivals. Crises in local, national, or global contexts are also a threat risk for tourism arrivals due to their reasons and potential results. Therefore, each crisis in the tourism industry, regardless of its size and content, should be taken into account and be managed properly. It is a fact that the tourism industry is very sensitive to crises, especially if there are perceived health and/or security concerns for visitors. Among a variety of different crises, pandemic conditions such as COVID-19 are also directly affect tourism demand, which is the core element for tourism activities. Hence, tourism demand shapes the destinations and tourism supply. Based on this reality, the general effects of the COVID-19 pandemic and its reflections on managerial operations in thermal hotels are discussed in this study. In the study, authors have also aimed to reveal the lessons learned from the COVID-19 for hotel managers.

Crisis management is preparation for low-probability or unexpected events that could threaten an organization's viability, reputation, or profitability (Pride & Ferrell, 2009: 455). Recently, crisis management has become more significant for tourism enterprises because several infectious diseases threaten the global economy and human lives (Shang, Li, & Zhang, 2021) and dependence on intangible knowledge capital to survive in an unsteady market that frequently witnesses economic turmoil (Paraskevas et al., 2013: 131). As a result, each hotel in the tourism industry develops a variety of managerial tools to prevent crises.

However, they cannot have a hold on macro crises, while they can prevent micro crises with those efficient tools. COVID-19 pandemic is a global crisis for hotels like every enterprise operating in economic markets. In other words, hotels have some barriers to cope with the COVID-19 pandemic. Even they behave proactively, some global restrictions such as flight bans and shutdowns of hotel operations force hotel managers to develop unprecedented methods and techniques. Hotel managers could turn the COVID-19 pandemic into an opportunity with those methods and techniques and could gain some advantages from this undesired atmosphere. Considering this fact, this study stresses the managerial advantages of the COVID-19 pandemic in terms of thermal tourism management with qualitative research.

2 Why Crises Matter in Tourism?

In Wuhan, Hubei Province, after a cluster of cases of pneumonia were reported by Wuhan Municipal Health Commission, China, on December 31st, 2019, World Health Organisation (WHO) reported on social media on January 4th that there was a cluster of pneumonia cases with no deaths in Wuhan. This statement was also identified as a novel coronavirus. The first COVID-19 case outside of China was detected in Thailand on January 13th. Later on, the novel coronavirus outbreak (2019-nCoV) was declared as a Public Health Emergency of International Concern (PHEIC) on January 30th by the advice of the Emergency Committee (EC). Deeply concerned by the alarming levels of transmission and severity, WHO assessed that COVID-19 could be considered as a pandemic (on March 11th) (WHO, 2020). On this date, the first case in Turkey was recorded and the first death due to COVID-19 occurred on 15 March. As of the beginning of March 2020, the rate of the epidemic slowed down in China, while there was an increase in COVID-19 cases and deaths due to this infection in Iran, the Republic of Korea (South Korea), and Italy. In the ongoing process, serious case increases were seen first in Europe and then in North America, and as of the beginning of May 2020, the epidemic continued to be seen in almost all countries in the world (Ministry of Health of Turkey, 2020).

Due to the transmissions of Covid-19, one of the firstly announced travel restrictions is came from North Korea on January 21, 2020 (Smith & Zhang, 2020). A few days later, nearly all of China's neighboring countries have restricted travel to and from the Republic of China (O'connor, 2020). The U.S. announced travel restrictions one month later from China's first warning about the pandemic (Bollyky & Nuzzo, 2020). In Turkey, it has decided that to stop all flights from China as part of coronavirus on February 5, 2020. All the passengers of far eastern countries like Japan, Taiwan, Thailand, Hong Kong, Singapore, South Korea, and Malaysia, were permitted entry to Turkish airports by screening with thermal cameras and considering coronavirus measures (Hürriyet Daily News, 2020). On March 13, 2020, The Ministry of Transport and Infrastructure, Turkey announced that flights to nine countries (Germany, France, Spain, Norway, Denmark, Belgium, Austria, Sweden, and the Netherlands) were stopped until April 17, 2020 (Keleş, 2020).

As the host country of the 2020 Summer Olympics, Japan closed its borders to all the non-residents until January 31, 2021; in addition, decided to organize Olympic Games without non-resident attendants (Wen, 2020). This causes a small-scale crisis because of sold tickets and refunds.

The coronavirus pandemic is considered a crisis that has affected the whole world in a short time and is unprecedented in terms of spread and impact. Although the state of panic that has been experienced all over the world since the beginning of the epidemic seems to have calmed down to some extent with the effect of vaccine studies, the virus is constantly mutating and the uncertainty regarding the future of the epidemic continues. Due to its structural characteristics, tourism is one of the sectors that are most affected by the epidemic and that have the highest share of this uncertainty. The coronavirus pandemic contains important clues for the answer to the question of why crises matter in tourism.

A crisis is a disruption that physically affects a system as a whole and addresses its core assumptions, subjective self and existential essence (Pauchand & Mitroff, 1992: 15 cited by Giousmpasoglu, Marinakou, & Zopiatis, 2021: 1298). Faulkner (2001: 136) evaluates the situations that negatively affect the functioning of the tourism industry in two groups as crisis and disaster. Crises are used to describe a situation where the root cause of an event is, to some extent, caused by problems such as incompetent management structures and practices or an inability to adapt to change. Disasters refer to situations where an enterprise is faced with sudden, unpredictable catastrophic changes over which it have little control. The researcher makes the difference between crisis and disaster according to the cause of the negative event, its scale, and the size of the event. Accordingly, he defines enterprise-based situations like a crisis, and natural and human-induced exogenous situations as disasters (Berbekova, Uysal, & Assaf, 2021: 1–2). The Covid 19 pandemic is a catastrophic situation that would be classified as a disaster according to Faulkner's definitions (Crespi-Cladera, Martin-Oliver & Pascual-Fuster, 2021: 2).

Tourism briefly expressed as "spatial movement of people" (Hu, Teichert, Deng, Liu, & Zhou, 2021) or "area of human activity" (Faulkner, 2001: 136), is one of the most important sectors of many countries in terms of economic and social aspects (Berbekova, Uysal, & Assaf, 2021: 1). Due to its structure, tourism can operate depending on basic elements such as peace, security, stagnation, free movement, and operational characteristics of sector enterprises (Giousmpasoglu, Marinakou, & Zopiatis, 2021: 1297). Peace, safety, and security are the primary conditions for the normal tourism development of a destination (Wang & Ritchie, 2012: 1057). Tourism is a complicated sector and vulnerable to adverse events due to the multitude of interdependencies and interactions among stakeholders (Berbekova, Uysal, & Assaf, 2021). Since tourism is not a basic need, the high sensitivity of tourism demand to economic, social, political, natural, and health (Pandemic)-based negativities results that the sector is one of the most vulnerable to all disasters and crises.

The closure and social distance policies implemented by governments for precautionary purposes when the Covid 19 pandemic emerged, directly affected tourism (Liu & Yang, 2021) because of the intensity of human mobility and close interaction. National and international travel restrictions, government shutdowns, and stay-at-home orders (Yang & Han, 2021: 1232) have caused the tourism sector to be one of the sectors most affected by the pandemic crisis.

3 Advantages of Disadvantages

Since crises and disasters are negative situations or events that occur suddenly or unexpectedly by their nature, they create disadvantageous situations by causing disruptions in the current functioning of businesses. In this direction, the covid 19 pandemic, which has obstacles such as travel restrictions, government shutdowns, and requires rules such as hygiene, distance, and reducing contact, has negatively affected tourism enterprises both in the process of stopping the activities and in the transition to normalization. The tourism statistics shared by World Tourism Organization for 2020 and 2021 are a clear indicator of the contraction in the sector. The main disadvantage of crises is the disruption of normal functioning. Crises adversely affect the normal functioning of activities at the macro and micro levels. In this context, the disadvantages of the Covid 19 pandemics in the tourism sector, especially in hotel businesses, can be listed as follows:

- Shrinking of demand,
- Changing the expectations of the demand,
- Decreased income of businesses,
- Increase in unemployment as a result of the reduction or dismissal of employed personnel,

- Decrease in the income of the suppliers of the sector in line with the decreasing demand, Businesses have to comply with some external obligations (government decisions, etc.) due to covid 19,
- The problem of adaptation of the operational activities of the enterprises to the new norms due to Covid 19,
- Additional costs to the business of the implementation of new norms due to Covid 19
- Increase in technology-based investment costs due to Covid 19,
- The deterioration of the financial structures of the enterprises,
- The inadequacy of businesses in the face of the need to innovate.

The spelling of the word crisis in the Chinese alphabet has two meanings: "danger" and "opportunity". This combination of the word actually explains the real character of the crisis phenomenon very well (Küçükaltan & Çitfçi, 2015: 252). In Webster's dictionary, the definition of crisis emphasizes the turn of events for the better or worse. Therefore, crises and disasters have transformational connotations with positive and negative consequences (Faulkner, 2001: 137). For this reason, although the crisis is seen as a negative situation or event by its nature, it is also an opportunity to take advantage of the disadvantageous situation experienced by a destination or enterprise at the macro and micro level. Thus, disadvantages caused by the crises mentioned above also refer to the situations and conditions that can be a means for enterprises to gain competitive advantage against their competitors in the sector with the right policies and strategy implementations.

When the results of the Covid 19 pandemic are evaluated in general; it is seen that it causes a great shrinkage in the sector, economic losses and a decrease in employment. While the pandemic obliges enterprises to adapt to new business processes and the new normal, it has compelled them to bear new costs and invest in technology during the reopening process. However, crisis atmosphere also contain opportunities that will enable enterprises to gain an advantageous position for the future. Because crises are opportunities to review marketing mix or market segmentation, to innovate in use of social media, downsizing, simplification of processes, workforce, skills and qualifications of workforce, and retraining. For example, Giousmpasoglu, Marinakou and Zopiatis (2021: 1312) stated that the new business environment created by the Covid 19 pandemic enables managers to identify new markets and segments and better communicate with stakeholders to increase revenues in difficult times. This new business environment also makes it easier to use technology that actually requires a challenge. Another point that the researchers drew attention to is the opportunity to recruit more talented people from the business pool full of talented people, which has grown due to the loss of jobs of many people due to the pandemic, to create work teams that are more sensitive to business processes, more prepared for crises, and highly adaptable to stress and challenges.

Faulkner (2001) emphasizes that crises create market shrinkage, but offer creativity and an opportunity to consider new options. In this context, businesses need to consider issues such as innovation and differentiation more carefully, both in terms of tourists' behavior and their own operational services. In this way, it is possible that the crisis contributes to their competitiveness. Bonn and Rundle-Thiele (2007: 619) states that crises offer organisations the opportunity to reconfigure their management structure in order to better cope with the shock events. With its consequences as a crisis, the Covid 19 pandemic forces managers to prioritize both making themselves better equipped and preparing their teams for adverse conditions adopting a more human-oriented paradigm in order to protect their businesses against potential crises, risks and vulnerabilities and to minimize the damage.

As a crisis, the Covid 19 with its consequences forces managers to consider as a priority both to improve themselves and to prepare their teams for adverse conditions in order to protect their businesses against potential crises, risks and vulnerabilities and to minimize the damage, and also to adopt a more people-oriented paradigm (especially people-oriented HRM strategy and employee training) (Giousmpasoglu, Marinakou, & Zopiatis, 2021: 1300). This is a situation that can be considered as an advantage in terms of protecting the competitive structure of enterprises and requires sector managers to gain a more holistic view of different situations and conditions and to think strategically and be more successful in future applications. Yang and Han (2021: 1243), from a different perspective, emphasizes that the Covid 19 forces hotels and food and beverage enterprises to learn operational processes such as takeaway, which they are not experts in, to develop new ways of using social media, and to focus on macro-level issues from daily activities. The authors also draw attention to the fact that successful applications in these matters create an advantage for the firms. One of the biggest problems of managers in hotels is that they cannot implement the strategy of being financially strong due to its cost. However, in crises such as Covid 19, one of the most basic determinants of the survival of a firm is to be financially strong. Being financially strong not only enables firms to survive in crises, but also provides managers with the opportunity to implement this strategy as it creates chances during and after the crisis. At this point, because firms with lower operational leverage (lower relevance of fixed costs) will be in a better position to survive the crisis (Crespi-Cladera, Martin-Oliver and

Pascual-Fuster, 2021: 12), the managers may be held this strategy as an opportunity. Bonn and Rundle-Thiele (2007: 619) emphasize that experiencing a crisis is a learning tool in terms of management, while Wang and Ritchie (2012: 1065) underline that such an experience increases managers' interest in crisis planning. In other words, as emphasized in the message intended to be given in the proverb *"One calamity is more valuable than a thousand counsels"*, it can be said that the crisis as a calamity leads the senior management to develop their proactive behavior skills, to develop a holistic perspective on how they can use all the resources they have correctly and how they can reconsider their relations with their stakeholders.

4 The Role of Managers in Crises

The main determinant in the continuation of the facilities in line with the business objectives in the tourism enterprises is the top management. Since tourism consists of many different sub-sectors, the characteristics of the sector enterprises and the originality of their services require different planning and management strategies (Wang & Ritchie, 2012: 1057). Research suggests that managers should have different roles and abilities in areas such as operational, managerial and leadership, humanitarian, information technology, financial and marketing management, depending on the conditions of being affected by factors such as situation and environment. Studies suggest different roles and abilities for hotel General Managers, which are affected by context and environment; such as operational skills, managerial and leadership skills, human skills, as well as, abilities in information technology, financial management and marketing (Giousmpasoglu, Marinakou, & Zopiatis, 2021: 1300). In this context, among the situational and environmental factors, especially the conditions caused by events such as crisis and disasters make the roles, abilities and perspectives of the managers important in terms of protecting the existence and competitiveness of the business. Since the tourism sector is, by its very nature, one of the sectors with the highest probability of experiencing crisis and disaster situations, crisis management is not an extra role, but rather a principle and necessity (Berbekova, Uysal, & Assaf, 2021). Ritchie (2004) states that the globalization of tourism increases the market share and profitability of the sector enterprises, but also makes the sector and enterprises open to global risks, and emphasizes that crisis and disaster management should be the main competence of tourism managers. Another point that the researcher draws attention to is that understanding the life cycle and potential effects of crises and disasters makes it easier to develop strategies.

In literature, there are different classifications about the stages of the crisis in the field of tourism. Faulkner (2001: 140) describes the steps of the crisis as pre-event, prodromal, emergency, intermediate, long term (recovery), resolution. Ritchie (2004: 680), who states that crisis management in tourism requires a strategic, holistic and proactive approach, classifies the steps as "(1) developing proactive scanning and planning; (2) implementing strategies when crises or disasters occur; and (3) evaluating the effectiveness of these strategies to ensure continual refinement of crisis management strategies" in this context. While Ritchie (2004) states that knowing the life cycle of crises and understanding their potential effects makes it easier to develop strategies, he also points out that due to the chaotic nature and uncertainty of crises and disasters, this approach is insufficient for managers and narrows their field of action. For this reason, crisis management should not be considered as researching and planning against every possible crisis but as developing the ability to react flexibly and to make the right decision instantly (Bonn & Rundle-Thiele, 2007: 619).

Covid 19 is the best example in this context. Because the Covid 19 pandemic has caused negative effects such as financial loss, job loss, and occupancy, which has not been caused by any crisis in the tourism sector so far, due to the conditions (banning air transportation, the inability of ships to leave the ports, forcing restaurants, hotels, and destinations to close to maintain social distance, etc.) it has created since it started (Giousmpasaoglu, Marinakou, & Zopiatis, 2021: 1298). Covid 19 has affected the travel behavior of tourists and service approach in the tourism sector and required tourism enterprises to continue their activities within the limitations of external rules and procedures (Kim & Pomirleanu, 2021: 2–3).

The Covid 19 pandemic, which continues to affect all sectors today, can be characterized by a high level of uncertainty and unpredictability of its consequences compared to previous disasters and crises. When the change caused by the Covid 19 pandemic, which is unknown how it will develop and appear in the future, due to the fact that it has not been brought under control yet, is observed, it is seen that it has created great changes in both travel behaviors and management activities from the beginning (Berbekova, Uysal, & Assaf, 2021). It is possible to collect the main issues that the Covid 19 pandemic has made more important in the execution of activities in the tourism sector and what managers should focus on in the continuation of business activities are demand (tourists), managerial practices, operational activities, and human resources management.

Demand (Tourists): Travel limitation and health concerns, one of the most important effects of the Covid 19, directly affected tourists, who constitute the demand of tourism enterprises. To protect themselves, people have stopped travels and tended towards minimal human interaction. In this respect, health and social interaction stand out as two main determinants on the demand side. These two determinants have led to new forms and changing expectations in people's daily life reorganization and touristic travel (Kim & Pomirleanu, 2021). Hu et al. (2021) state that the Covid 19 pandemic has directed people to cocooing behavior, and they describe cocooning behavior in the hospitality industry as the behavior in which tourists escape from daily life and seek a quiet atmosphere (camping areas, escaping from social contact, etc.). Other points that the researchers drew attention to the change caused by the pandemic in the perspectives of tourists about hotel services are the restriction of the use of general hotel services due to the social distance rule, and the priority of hygiene conditions in the rooms. In other words, the pandemic has shifted the expectations of tourists from hedonic to utilitarian hotel features.

Managerial Practices: Crises, disasters and shock events cannot be stopped, but their effects can be limited with managerial practices. For this reason, managers should make vital strategic and operational decisions to reduce the effects of shock events such as crises. In this context, Bonn and Rundle-Thiele (2007: 616–619) emphasize that it is important for managers to make quick and correct decisions in their managerial practices, to follow strategies to reduce organizational stress, and to act with more intuition than analysis. Emphasizing that the short-term planning of crisis management practices in epidemic-like crises is not suitable for the covid 19 pandemic, Lai and Wong (2020) draws attention to the fact that the strategies of hotel managers to increase demand by reducing prices with the expectation that the epidemic will end in a short time are not a correct managerial practice. From a managerial point of view, the recommendations of the researchers for human resources are to reduce the working time and day in the pre-crisis phase and to reduce the workforce during the crisis. Outsourcing and re-employment are not appropriate managerial practices for conditions such as the Covid 19 pandemic. In order to reduce labor and operating costs (without impairing the service quality), limiting service provision (especially for the common use areas of tourists) and training employees should be considered among the managerial practices that should be applied during the crisis. Jiang and Wen (2020: 2568) point out that physical and mental well-being has become important issues in the expectations of tourists during the Covid 19 period, and they draw attention to the fact that services such as meditation programs, digital detox, fitness programs, healthy diet programs should be considered among the managerial strategies.

Operational Practices: Deloitte (2020: 40) categorized the industries into four segments based on the degree of change in product offerings and the impact of COVID-19 on business capabilities: the change hero, the wave surfer, the restructurer and the redefiner. Hotels are included in the "restructuring" group. Because a major part of the service delivery areas in hotels is the common use areas of the guests. Therefore, hotels are the businesses with the highest risk for health and infection issues, which people focus on most due to the pandemic. This highlights that hygiene should be a priority in the business model to ensure tourists have confidence that they are protected by the hotel (cited by Denizci-Guliette & Chu, 2021: 605). In addition, the fact that the new market requirements (especially the expectations of the demand) that emerged with the Covid 19 pandemic have made less contact the basic norm, has increased the use of new technologies that make this possible (Giousmpasoglu, Marinakou, & Zopiatis, 2021: 1313). Accordingly, the use of artificial intelligence (AI) to reduce or eliminate human to human interaction in operational areas in hotels, and self-service technologies (digital switch, service robots, self-service kiosks, smart speakers, drones, etc.) have become the main elements of service innovation motivation with the pandemic (Jiang & Wen, 2020; Lui & Yang, 2021).

Human Resources Management: The first priority of hotel management is to provide services flawless that meet the expectations of tourists and reinforce their sense of trust. In the coronavirus pandemic; social distance, hygiene, cleanliness, etc. are the main determinants of tourists' expectations from hotels (Hu et al., 2021: 8). This makes human resources crucial as a substantial determinant of service encounter and makes employee training vital so that they can implement new service forms and new rules and practices in service encounter during the pandemic (Lai & Wong, 2020: 3150). As the Covid 19 pandemic caused a decrease in the demand for hotels due to shutdowns and travel restrictions, hotel managers were forced to reduce their employees because of the demand insufficiency. In addition, the risk of contamination has caused the situation of working in a stressful work environment, especially in the departments that are in close contact with guests. Thus, in terms of human resource management, implementing effective strategies that reduce the stress of employees has become important (Bonn & Rundle-Thiele, 2007: 619).

Covid-19 with the new environment created by it has completely changed the general operations/status quo in tourism enterprises. This fact is also true to the roles and responsibilities of hotel managers. For instance, continuously changing customer expectations and behaviors require implementing new business analyses. As a result, in order to maintain their competitive structure, businesses have started to run to value-based pricing instead of demand-based pricing. Hence, previous data and budgets are not sufficient for revenue management in this crisis. So, in marketing efforts hygiene conditions, customer health and corporate social responsibility applications have become more important. Further, cooperation among departments, the continuation of activities with joint commissions in distribution management, implementing short-term strategies in terms of revenues can be counted as some of the priority issues in times of crisis (Denizci-Guillet & Chu, 2001).

Conclusion

As it is well known, chaos and change are important parts of management. Undoubtedly, tourism due to its sectoral nature is one of the most affected sectors from chaos environment, which is expressed as crisis and disaster. Thus chaos environment should be taken into account in tourism management (Ritchie, 2004). In this context, the Covid 19 pandemic has caused a sudden contraction all over the world, both in general and in tourism, unlike the crises experienced so far, with the necessity of social distance, travel restrictions and stay-at-home compulsions (Denizci-Guillet & Chu, 2001). For this reason, the Covid-19 pandemic has made it necessary to develop new policies, strategies and practices in terms of management. When the changes in the tourism sector and academic studies are examined from the beginning of Covid-19 up to today, it is observed that there have been great changes in the fields of demand (tourists), managerial practices, operational activities, human resources management and the status quo. The Covid-19 pandemic has made it mandatory for the industry to restructure its service approach, especially as it affects the travel behavior of tourists (Kim & Pomirleanu, 2021). In this new order shaped by Covid-19 pandemic and taught as crisis some issues have come into forefront such as,

- The increasing importance of self-service technologies,
- Awareness about the importance of being financially strong,
- The emergence of social media in marketing activities,
- The change in the focus of messages in marketing activities,
- The understanding of the importance of close cooperation among stakeholders,
- The understanding of the synchronized action among different departments within the enterprise,
- The importance of employee training.

One of the positive aspects of the Covid 19 pandemic in terms of management is that it reveals that the importance of crisis management, and crisis management as one of the vital parts of management practices should be constantly considered. Furthermore, the adoption and demand of technology-based innovation practices, which were previously seen as a weakness in service quality by tourists, as a standard in service delivery in the new order, can be considered as another positive side of Covid-19.

Bibliography

Berbekova, A. Uysal, M. and Assaf, A.G. (2021), A Thematic Analysis of Crisis Management in Tourism: A Theoretical Perspective, *Tourism Management*, 86, pp. 1–13.

Bollyky, T. J. and Nuzzo, J. B. (2020), Trump's 'early' travel 'bans' weren't early, weren't bans and didn't work, https://www.washingtonpost.com/outlook/2020/10/01/debate-early-travel-bans-china/ (Access Date: July 10, 2021).

Bonn, I. and Rundle-Thiele, S. (2007), Do or Die-Strategic Decision-Making Following a Shock Event, *Tourism Management*, 28, pp. 615–620.

Crespi-Cladera, R. Martin-Oliver, A. and Pascual-Fuster, B. (2021), Financial Distress in the Hospitality Industry During the Covid-19 Disaster, *Tourism Management*, 85, pp. 1–13.

Deloitte (2020), Impact of the COVID-19 crisis on short and medium-term consumer behavior: Will the COVID-19 crisis have a lasting effect on consumption?, https://www2.deloitte.com/ (Access Date: September 30, 2021).

Denizci-Guillet, B. and Chu, A. M. C. (2021), Managing Hotel Revenue amid the COVID-19 Crisis, *International Journal of Contemporary Hospitality Management*, 33 (2), pp. 604–627.

Faulkner, B. (2001), Toward a Framework for Tourism Disaster Management, *Tourism Management*, 22, pp. 135–147.

Giousmpasoglou, C. Marinakou, E. and Zopiatis, A. (2021), Hospitality Managers in Turbulent Times: The COVID-19 Crisis, *International Journal of Contemporary Hospitality*, 33 (4), pp. 1297–1318.

Hu, F. Teichert, T. Deng, S. Liu, Y. and Zhou, G. (2021), Dealing with Pandemics: An Investigation of the Effects of COVID-19 on Customers' Evaluations of Hospitality Services, *Tourism Management*, 85, pp. 1–14.

Hürriyet Daily News, (2020), *Turkey stops all flights from China as part of coronavirus measures*, https://www.hurriyetdailynews.com/turkey-to-suspend-flights-from-china-until-end-of-month-151705 (Access Date: July 10, 2021).

Jiang, Y. and Wen, J. (2020), Effects of COVID-19 on Hotel Marketing and Management: A Perspective Article, *International Journal of Contemporary Hospitality Management*, 32 (8), pp. 2563–2573.

Keleş, N. (2020), *Transport and Infrastructure Minister Turhan: Flights to 9 Countries Were Stopped*, https://www.aa.com.tr/ (Access Date: July 10, 2021).

Kim, E. J. and Pomirleanu, N. (2021), Effective Redesign Strategies for Tourism Management in a Crisis Context: A Theory-in-use Approach, *Tourism Management*, 87, pp. 1–12.

Küçükaltan, D. and Çiftçi, G. (2015), Turizm İşletmelerinde Kriz Yönetimi, İçinde O. Akova, İ, Kızılırmak, H. Tanrıverdi (Eds.), *Turizm İşletmeciliği Temel Kavramlar ve Uygulamalar* (pp. 251–271). Ankara: Detay Yayıncılık.

Lai, I. K. W. and Wong, J. W. C. (2020), Comparing Crisis Management Practices in the Hotel Industry Between Initial and Pandemic Stages of COVID-19, *International Journal of Contemporary Hospitality Management*, 32 (10), pp. 3135–3156.

Liu, C. and Yang, J. (2021), How Hotels Adjust Technology-Based Strategy to Respond to COVID-19 and Gain Competitive Productivity (CP): Strategic Management Process and Dynamic Capabilities, *International Journal of Contemporary Hospitality Management*, 33 (9), 2907–2931.

Ministry of Health of Turkey, (2020), COVID-*19 (SARS-CoV-2 infection)*, https://covid19.saglik.gov.tr/ (Access Date: July 10, 2021).

O'connor, T. (2020), *China's Neighbors Close Borders As Country's Coronavirus Cases Surpass Last Major Outbreak*, https://www.newsweek.com/china-neighbors-close-borders (Access Date: July 10, 2021).

Paraskevas, A. Altinay, L. McLean, J. and Cooper, C. (2013), Crisis Knowledge in Tourism: Types, Flows and Governance, *Annals of Tourism Research*, 41, pp. 130–152.

Pauchant, T. C. and Mitroff, I. (1992), *Transforming the Crisis-Prone Organization: Preventing Individual, Organizational, and Environmental Tragedies*. New Jersey: Jossey-Bass.

Pride, W. and Ferrell, O. C. (2009), *Foundations of Marketing* (3rd Ed.). Boston: Houghton Mifflin.

Ritchie, B. W. (2004), Chaos, Crises and Disasters: A Strategic Approach to Crisis Management in the Tourism Industry, *Tourism Management*, 25, pp. 669–683.

Shang, Y. Li, H. and Zhang, R. (2021), Effects of Pandemic Outbreak on Economies: Evidence from Business History Context. *Frontiers in Public Health*, 9, https://doi.org/10.3389/fpubh.2021.632043 (Access Date: December 7, 2021).

Smith, J. and Zhang, L. (2020), *North Korea Suspends Foreign Tourism over Coronavirus Fears: Tour Companies*, https://www.reuters.com/ (Access Date: July 10, 2021).

Swarbrooke, J. and Horner, S. (2007), *Consumer Behaviour in Tourism* (2nd Ed.). London: Butterworth-Heinemann.

Wang, J. and Ritchie, B. W. (2012), Understanding Accommodation Managers' Crisis Planning Intention: An Application of the Theory of Planned Behaviour, *Tourism Management*, 33, pp. 1057–1067.

Wen, A. (2020), *Japan Closes Borders, Limits Travel to Citizens after New COVID-19 Strain from UK Emerges Spread Rapidly via Human-to-Human Transmission*, https://news.yahoo.com/japan-closes-borders (Access Date: July 10, 2021).

WHO (2020), *Rolling Updates on Coronavirus disease (Covid 19)*, https://www.who.int/emergencies/diseases/novel-coronavirus-2019/events-as-they-happen (Access Date: December 7, 2021).

Yang, M. and Han, C. (2021), Revealing Industry Challenge and Business Response to Covid-19: A Text Mining Approach, *International Journal of Contemporary Hospitality Management*, 33 (4), pp. 1230–1248.

Bayram ŞAHIN, Selda GÜVEN and İbrahim MISIR

Crisis, Terrorism, Epidemic Diseases and Their Impacts on International Destination Selection

1 Introduction

First step for understanding tourist behavior is destination selection. Tourists select the destination that they will visit as per various internal and external factors. One of these factors is tourists' perception of safety and security with regards to the destination. When this perception is negative, this causes tourists to delay or cancel their holiday plans and turn to places where they feel safer and this particular may cause major crises in relation to destination. Terrorism and epidemics can be listed among the particulars that negatively affect the perception of the destination. In this part, primarily destination selection, safety and security in tourism will be discussed. Then, information will be given about subjects of crises, crisis types, life cycle of crises, tourism and crisis, terrorism, tourism and terrorism, epidemic diseases, tourism and epidemic diseases, respectively.

2 Selection of Destination

Destinations can be defined as mixtures of tourism products which offer an integrated experience to consumers, being geographically well defined such as a country, island or town, which can be subjectively interpreted by tourists depending on their itinerary, cultural background, purpose of visit, education level and past experience (Buhalis, 2000: 1). Destination selection has a more complex structure with respect to other products since it is an all-inclusive experience that requires the tourist to be away from the place where they are permanently located, that is related with a completely foreign environment, a different language, local culture and acceptable behavior in this culture (Fuchs & Reichel, 2011: 272). Destination selection is based on the process of negotiation between the tourist's characteristics and the destination's attributes. In the scope of risk, tourists make their destination choices according to their individual perceptions of travel risk (Karl, Muskat, & Ritchie, 2020: 2). Perceived risk in tourism can be defined as travel-related risks that affect the decision-making process of the tourist (Dolnicar, 2005: 205). Guillet, Lee, Law and Leung (2011: 557) state that the main element of understanding tourist behavior is destination selection. At this point, tourist behavior and destination selection as a result of this behavior are shaped by the tourist's perception of security and safety, and are inevitably linked to these issues (Hall, Timothy, & Duval, 2009: 2; Micić, Denda, & Popescu, 2019: 39).

Tourists often look for a relaxing and carefree holiday and are hence susceptible to violence in resorts. Ironically, for most of human history, traveling has been correlated with the traveler's physical integrity, perception of risk, and fear (Neumayer, 2004: 259). Terrorist acts damage the security and attractiveness image of a destination and endanger the existing tourism industry in the destination (Pizzam & Smith, 2000: 123). An important success factor for tourism destinations is the ability to provide a safe, predictable and secure environment for the visitor (Speakman & Sharpley, 2012: 67). From this respect, it can be stated that physical security is the most important particular for tourists, and that the destination's being safe and far from actions of violence and terrorism can positively influence the destination choice of tourists (Ali, Shah, & Khan, 2018: 130). When we consider destinations with regards to terms of epidemics, it can be stated that new epidemics may occur, tourism, travels or destinations may change completely, individuals may keep away from traveling or be prevented from traveling due to life safety. In fact, the epidemic that is still experienced today could not be kept under control despite various measures and the future of tourism is quite uncertain. As Yeoman (2011: 5) states, the terrorist attacks in 2001 changed the perspective on security in tourism. A similar conclusion can be made for epidemics. In other way of saying, epidemics that have been experienced and are likely to occur in the future have the potential to cause radical changes in relation to the perspective of tourism, travel and destinations. Rittichainuwat and Chakraborty (2009: 416) state that both terrorist actions and epidemics increase the risk perception by reducing the confidence of tourists in international travel and damaging the destination. Furthermore, price reductions applied to increase demand negatively affect the image of the destination rather than motivating tourists.

Taking the dependence of tourism destinations on tourism-related activities into consideration, their vulnerability to crisis formation increases significantly when it is considered that they must have a positive image for success and maintain this image (Santana, 2004: 300). Crises are turning points for destinations. They may even include positive results such as innovation, recognition of new markets. Following the crisis, there may be situations such as learning from crises, making policy changes, adapting and changing strategies that do not work effectively (Blackman & Ritchie, 2008: 46). However negative outcomes of crises must not be neglected. Hence, statement of Racherla and Hu (2009: 562) specifying that crises give damage to image of destination as it requires many years to restore supports this expression. Similarly, Ertaş, Sel, Kırlar-Can, and Tütüncü (2021: 1493) stated that crises have a negative impact on the reputation, marketability and perception of the destination. There are different opinions about the effects of crises on tourists' destination choice. Farmaki (2021: 2) states that in the event of a crisis, tourists can cancel their trips, postpone them, head to an alternative destination or visit the destination where the crisis occurred, by undertaking all kinds of risks. According to Seabra, Reis, and Abrantes (2020: 1), tourists tend to choose the lower risk option, which gives them more potential gains. Tourism is more affected by crises compared to other industries. Because the main purpose of going on vacation is hedonism (Micić, Denda, & Popescu, 2019: 39). Failure to address potential tourists' safety and security concerns can create serious administration problems that can have long-term, devastating economic effects on the destination (Sirakaya, Sheppard, & McLellan, 1997: 2).

3 Safety and Security in Tourism

As it is the case in human life, the need for security also has a very important place in Maslow's hierarchy of needs. According to the theory, individual needs to satisfy that requirement adequately to complete his basic needs and move on to the next stage (Wahba & Bridwell, 1976: 213–214; Adler, 1977: 444; Lester et al., 1983: 83). In other way of saying, the individual will feel the need for security after satisfying his physiological needs and will not be able to go on to the next stage without satisfying this need (Mathes, 1981: 69). According to Korstanje and Tarlow (2012: 27), only living being among all living creatures having fear of death is humans, which leads people to more sheltered areas and creates a need for security. In this regard, it can be stated that the concept of security, which means protection against dangers and risks arising from deliberate actions (Mekinc & Cvikl, 2013: 39), is an important particular for all living things. From this perspective, the need for security is an innate feature of human nature (Kozak, Crotts & Law, 2007: 233), and security concerns may put tourists in a dilemma about the destination they will choose (Amir, Ismail, & See, 2015: 124). At this point, it can be stated that terrorist acts and epidemics seriously threaten the safety and lives of individuals, and that violence and health-based crises cause situations that are difficult and in some cases impossible to remedy in tourism.

It can be specified that the concerns about safety and security in tourism go back to the Ancient Greeks, and wars were suspended during the Olympic Games. It is stated that in the Middle Ages, safety and security concerns caused a

Tab. 1: The Travel and Tourism Competiveness Report 2019 Safety and Security

Rank	Country	Score (1–7)	Rank	Country	Score (1–7)	Rank	Country	Score (1–7)
1	Finland	6,7	48	Jordan	5,7	95	North Macedonia	5,2
2	Iceland	6,5	49	Chile	5,7	96	Kyrgyz Republic	5,2
3	Oman	6,5	50	Gambia	5,7	97	Tanzana	5,2
4	Switzerland	6,4	51	France	5,7	98	Russian Federation	5,1
5	Hong Kong (SAR)	6,4	52	Belgium	5,7	99	Argentina	5,1
6	Singapore	6,4	53	Algeria	5,6	100	Sierra Leone	5,1
7	United Arap Emirates	6,3	54	Nicaragua	5,6	101	Ethiopia	5,1
8	Luxembourg	6,3	55	United States	5,6	102	Cambodia	5,1
9	Portugal	6,3	56	Poland	5,6	103	Namibia	5
10	New Zeland	6,3	57	Slovak Republic	5,6	104	Angola	5
11	Qatar	6,3	58	Viet Nam	5,6	105	Bangladesh	4,9
12	Estonya	6,2	59	China	5,6	106	Paraguay	4,9
13	Japan	6,2	60	Tajikistan	5,6	107	Ukraine	4,8
14	Austria	6,2	61	Greece	5,6	108	Burkina Faso	4,8
15	Slovenia	6,1	62	Mongolia	5,6	109	Haiti	4,8
16	Spain	6,1	63	Kazakhstan	5,6	110	Burundi	4,8
17	Norway	6,1	64	Montenegro	5,6	111	Thailand	4,8
18	Czech Republic	6,1	65	Mauritania	5,6	112	Egypt	4,8
19	Australia	6,1	66	Israel	5,5	113	Lebanon	4,8
20	Brunei Darussalam	6,1	67	Moldova	5,5	114	Dominican Republic	4,7
21	Canada	6,1	68	Ghana	5,5	115	Mozambique	4,7
22	Malta	6	69	Italy	5,5	116	Uganda	4,7
23	Saudi Arabia	6	70	Eswatini	5,5	117	Cameroon	4,7
24	Netherlands	6	71	Serbia	5,5	118	Peru	4,7
25	Georgia	6	72	Benin	5,4	119	Kenya	4,6
26	Taiwan, China	6	73	Malawi	5,4	120	Guinea	4,6
27	Ireland	6	74	Iran	5,4	121	Cote d'Ivoire	4,6
28	Morocco	6	75	Costa Rica	5,4	122	India	4,5
29	Romania	6	76	Bosnia and Herzegovina	5,4	123	Congo D. Rep.	4,4
30	Korea Rep.	5,9	77	Zimbabwe	5,4	124	Brazil	4,3
31	Rwanda	5,9	78	Sri Lanka	5,4	125	Turkey	4,3

Tab. 1: Continued

The Travel and Tourism Competiveness Report 2019 Safety and Security							
32	Bahrain	5,9	79	Lesotho	5,4	126	Mexico	4,2
33	Cyprus	5,9	80	Indonesia	5,3	127	Chad	4,2
34	Malaysia	5,9	81	Zambia	5,3	128	Mali	4
35	Croatia	5,9	82	Liberia	5,3	129	Guatemala	4
36	Sweden	5,9	83	Lao PDR	5,3	130	Trinidad and Tobago	3,9
37	Lithuania	5,9	84	Panama	5,3	131	Jamaika	3,9
38	Azerbaijan	5,9	85	Uruguay	5,3	132	South Africa	3,9
39	Hungary	5,8	86	Senegal	5,3	133	Colombia	3,8
40	Armenia	5,8	87	Bostwana	5,3	134	Pakistan	3,7
41	Germany	5,8	88	Ecuador	5,2	135	Philippines	3,6
42	Latvia	5,8	89	Bolivia	5,2	136	Honduras	3,6
43	Denmark	5,8	90	Tunusia	5,2	137	Venezuela	3,3
44	Kuwait	5,8	91	Nepar	5,2	138	Yemen	3,2
45	United Kingdom	5,8	92	Cape Verde	5,2	139	Nigera	3,1
46	Mauritius	5,8	93	Bulgaria	5,2	140	El Salvador	3
47	Albania	5,8	94	Seychelles	5,2	141		

Source: World Economic Forum

decrease in travel (Kôvári & Zimányi, 2011: 59). By the 1950s, security and safety in mass tourism came to the forefront by changing its shape. Elements that determine this process of change are the spread of tourism to large masses, its covering more countries in the world as an economic development strategy, and the developments in the field of transportation (Ayob & Masron, 2014: 2). In the late 1960s and early 1970s, the increase in mobility together with the developments in public commercial transport revealed significant changes in globalization processes and hyper-mobility, which means huge growth in temporary mobility in some societies (Hall, 2010: 402). Even though the tourism industry has been severely damaged by crime, terrorism, food security, health concerns and natural disasters in the recent past, events of September 11 became a turning point in which such crises gained greater visibility (Breda & Costa, 2006: 187–188). In other words, September 11 can be stated as a milestone in which security and safety perceptions in tourism are shaped again. In fact, events such as crime, terrorism, food safety, health concerns and natural disasters have local, regional and global repercussions, causing tourism crises at the institutional, industrial and destination level (Henderson, 2007: 1). Today, ensuring the safety and security of tourists has become one of the basic elements of tourism (Liu, Pennington-Gray, & Krieger, 2016: 311). In addition, safety and security come to the fore in the competitiveness of the destination. From this perspective, it can be stated that one of the subtitles of the Travel and Tourism Competitiveness Report is safety and security, and that the countries included in the report are ranked according to scores ranging from 1 to 7 (Tab. 1). According to the table, Finland, Iceland, Oman, Switzerland and Hong Kong are the top five countries with the highest scores in the safety and security category, respectively. The countries in the last five in the list are Honduras, Venezuela, Yemen, Nigeria and El Salvador, from high to low. It is reported that Honduras is one of the poorest countries in Central America, a lot of crime is committed in the country, and 60 out of a hundred thousand people are murdered every year (Anadolu Agency, 2021). In Venezuela, on the other hand, it is stated that violence is a structural problem, and the government has made agreements with criminal organizations, reducing or completely removing the legal pressure and sanction power on the organizations (Voice of America, 2021). Yemen is currently experiencing the heaviest human tragedy in the world, it is estimated that more than a hundred thousand people have lost their lives in the last six years due to war, scarcity and health problems (Deutsche Well, 2021). Similarly, Nigeria and El Salvador are frequently on the world agenda with crises such as terrorism, gang wars, political instability, and health problems.

Crises such as terrorist acts, civil wars, epidemics and natural disasters have negative effects and lasting consequences on the travel and tourism industry (Brondoni, 2016: 9). This situation is an indicator of the vulnerability of tourism destinations (Baker & Coulter, 2007: 249). Essentially, it can be stated that the tourism industry is quite open to risks and prone to crises due to external events. Unlike internal events that can be evaluated and controlled by managers, external events are beyond their control and inherently carry more risk and uncertainty. Indeed, the unique characteristics of service-based industries potentially make it difficult to manage crises (Evans & Elphick, 2005: 135). The most effective fear that crises can create for a tourist is that he will never get to his destination or never return home alive. If the tourist is convinced of this point, serious decreases will occur in the demand for the destination (Norton, 1987: 31). Tourists are generally risk-averse. Therefore, when any real or perceived threat to their health and safety occurs, this may affect their decision to visit a particular destination (Speakman & Sharpley, 2012: 67). Risk and security perceptions affect the destination image and choice, causing permanent negative images that risks are associated with certain destinations (Sönmez & Graefe, 1998: 173). Even though at first stage it seems as if what affects the destination is crisis itself, it is the image of the destination that is really affected by the consequences of the crisis (Hannam, 2004: 259). The factor that determines the success or failure of a destination rather than other economic activities is the protection of tourists visiting the destination from all kinds of risks and providing safe and secure environments for tourists in the face of crises such as natural disasters, epidemics, political instability, armed conflict and terrorism (Volo, 2008: 84). When faced with a crisis in the tourism industry, it is important that both the public and private sectors respond quickly and effectively to the crisis (Li, Blake & Cooper, 2010: 450). In the graphic below (Graph. 1), the elements perceived as threats in the society are listed. Accordingly, the highest rate is infectious diseases with 58 %. It is not surprising that such a result came out in a study conducted in 2020. After all, the whole world is suffering from the Covid-19 pandemic. Other perceived threats are livelihood crises (55.10 %), extreme weather events (52.10 %), cybersecurity threats (39 %), digital inequality (38.30 %), prolonged stagnation (38.30 %), terrorist attacks (37.80 %), disillusioned youth caused by inequality of opportunity (36.40 %), social cohesion erosion (35.60 %), and human-induced environmental damage (35.60 %).

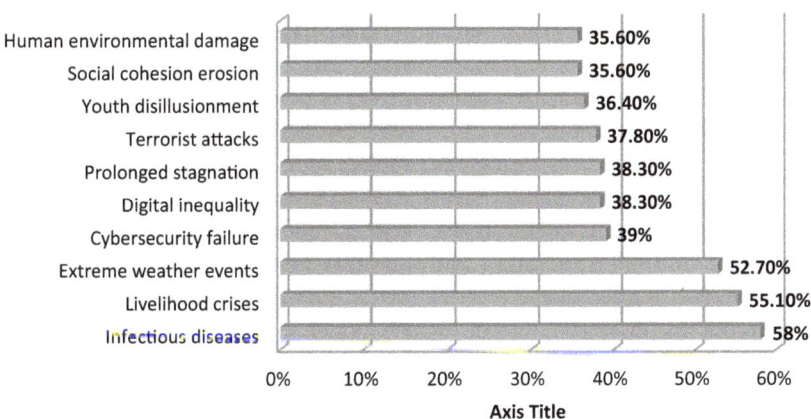

Graph. 1: Biggest Perceived Risk Threats to Global Society
Source: Statista, 2021

4 Concept and Types of Crisis

Concept of crisis, originating from the Greek word "krisis", means judgment, choice or decision (Maditinos & Vassiliadis, 2008: 68). Although there is no generally accepted definition (Ghaderi, Som, & Wang, 2014: 631), a crisis can be expressed as an unstable and important time period or situation that makes a difference in determining the outcome either as good or bad (Adnan, 2014: 162). Dyson and Hart (2013: 2), on the other hand, define crises as events that are interpreted as a serious threat to basic structures, values and norms, require urgent intervention, and create highly uncertain conditions. In general, a crisis is an unexpected and negatively oriented event that has the potential to lead to major negative consequences (Hong, Huang, & Li, 2012: 537). From an organizational perspective, a crisis can be defined as any action or failure of action that has a personal impact, which is perceived as harmful by the majority of its employees, customers or other stakeholders, that interferes with the ongoing functions, acceptable achievement, viability, and survival of the organization (Faulkner, 2001: 136). The nature of the crisis can result in shock, confusion, guilt, denial and anger, leading to action or inaction. The inexperience in dealing with such events reveals feelings of uncertainty and the expectation of various stakeholders to control crises and reallocate security and trust (Doepel, 1991: 177).

A crisis is a serious event that may have many causes, such as natural, political, financial or technical factors, that disrupts pre-existing functioning. After the crises, the stakeholders affected by the crisis have a widespread expectation that the old normal will be restored. However, after a serious crisis, it is generally accepted that a new work ethic and new ways of doing business are established (Laws, Prideaux, & Chon, 2007: 6). Crises can be classified as sudden crises where there is little or no warning, the problem cannot be investigated, or a plan cannot be drawn up; escalating crises that develop slowly, can be stopped or limited and persistent crises that can last for weeks, months or even years (Ritchie, 2004: 671). Mitroff, Shrivastava and Udwadia (1987: 286) first divided the crises into two as internal and external crises, and then addressed them as technical, economic, human-induced, organizational and social crises. Sausmarez (2007: 703) and Li, Blake and Cooper, 2010: 435 divide crises into natural crises such as hurricanes and earthquakes, and human-induced crises such as terrorist incidents and plane crashes. Tse (2006: 30) discussed the crises in tourism under four headings being natural disasters, civil conflicts, epidemics and technological errors. Crises, whether human-based or natural, have negative effects on tourism (Santana, 2004: 300; Li, Blake, & Cooper, 2010: 435; Perles-Ribes,

Ramón-Rodríguez, Rubia-Serrano, & Moreno-Izquierdo, 2016: 1215; Aliperti et al., 2019: 1).

4.1 Life Cycle of Crises and Crisis Management Approaches

It can be stated that in the literature, perspectives toward life cycle of crisis are in different forms as three-stage, four-stage, five-stage (Jing, Deng, & Zhao, 2012: 178; Stewart & Wilson 2016: 642) and six-stage (Jing, Deng, & Zhao, 2012: 178) ones. Accordingly, the three-stage model can be divided into three stages as pre-crisis, crisis stage and post-crisis (Li, Blake, & Cooper, 2010: 436; Jia, Shi, Jia, & Li, 2012: 139; Coombs & Laufe, 2018: 199; Tokakis, Polychroniou, & Boustras, 2019: 38). Howell and Miller (2006: 2); Kash and Darling, (1998: 181); Xiu-hua, Zhi-yong, and Lu-qi (2011: 346); Cushnahan (2014: 327) evaluates the stages of the crisis under four headings as pre-symptom, acute, chronic and recovery stage. In the pre-symptom stage, which is the first stage of the crisis, there are some symptoms and warnings about the crisis. Timely and appropriate intervention can prevent the crisis or mitigate its consequences. At this stage, it is easier to manage crises than other stages (Kash & Darling, 1998: 181) and it can also be expressed as the pre-crisis stage (Jaques, 2007: 154). In the acute crisis phase, the crisis really emerges and situations that trigger the crisis and cause damage occur (Stewart & Wilson, 2016: 642). The chronic phase of the crisis is also called the cleansing phase. In this period, recovery started and strengths and weaknesses emerged for future crises (Keown-McMullan, 1997: 6). The chronic stage refers to the lasting effects of the crisis. Individual crises can occur quickly, the lasting effects of the event can extend the life cycle of the crisis (Boudreaux, 2005: 11). One of the biggest challenges in managing a stage four crisis is the speed with which losses are intensified. In a successful crisis management, it is ensured that all relevant parties are aware of what is going on, and damage and needs assessments are made so that everyone can take part in the crisis response (Liou, 2015: 252). During the recovery phase, the crisis is handled internally and externally (Hale, Dulek, & Hale, 2005: 113), it consists of post-crisis activities related to assessment and documentation (Penrose, 2000: 162). The five-stage crisis lifecycle consists of stimulus detection, preparedness/prevention, damage limitation, recovery, and learning (Mitroff, Shrivastava, & Udwadia, 1987: 284; Bhaduri, 2019: 538). Augustine (1995), on the other hand, in his work titled "Managing Crisis You Tried the Prevent" in the Harvard Business Review in 1995, divides the stages of the crisis into six levels as avoiding the crisis, preparing for the crisis, confirming the crisis, controlling the crisis, resolving the crisis, and making a profit from the crisis. Pedersen, Ritter, and Benedetto (2020: 315) classify the crisis in a five-stage process as normality before the crisis, emergence, formation, outcome and post-crisis normality.

Crisis management is a process and management model applied in extraordinary situations. It includes certain activities such as perception of crisis signals, prevention of possible negative effects, survival with minimum loss, implementation and control of recovery preparation activities (Sahin, Ulubeyli, & Kazaza, 2015: 2299). Crisis management is defined as a systematic process supported by both internal and external stakeholders, including identifying the signals of crisis, preventing and preparing for possible damages, getting rid of the crisis and learning lessons from the crisis (Hong, Huang, & Li, 2012: 537). Approaches to crisis management can be classified as proactive and reactive. While the proactive approach refers to activities such as mitigation and preparedness that are planned and carried out before the crisis occurs, the intervention and recovery activities carried out during and after the crisis represent the reactive approach (Mojtahedi & Oo, 2017: 40). In the proactive stage, a potential crisis is noticed, attempts are made to avoid the problem completely or at least to minimize its consequences (Pforr & Hosie, 2008: 255). Reactive crisis management, on the other hand, focuses on currently occurring, latent or acute crises (Martens, Feldesz, & Merten, 2016: 91). Loosemore (1998: 140) states that reactive practices towards crisis are generally associated with weak management practices, and reactive practices are required in order to reduce the impact of crises in certain environments. As crises are complex, multidimensional, not well defined and highly intercorrelated problem systems (Alpaslan & Mitroff, 2021: 1), they require the use of proactive and reactive strategies together (Hernantes, Rich, Laugé, Labaka, & Sarriegi, 2013: 1749).

4.2 Tourism and Crisis

When considered in terms of tourism, the area where the crisis is effective is the destination and the whole country where the destination is located (Ghaderi, Som, & Henderson, 2012: 80). In the context of tourism destination, crisis can be expressed as events that threaten the normal functioning and behavior of tourism enterprises. The crises that occur negatively affect the perceptions of the visitors about that destination, the image of the destination, cause a decline in the local travel and tourism economy, and interrupt the continuity of commercial activities for the local travel and tourism industry with the decrease in tourist arrivals and expenditures (Sönmez, Apostolopoulos, & Tarlow, 1999: 13–14). In other words, the dependency relationship between the components that make up tourism shows that a crisis can spread to another area, and the sudden decrease in tourist arrivals affects domestic and foreign government institutions, tour operators and travel agencies together with accommodation, attraction and transportation providers (Henderson, 2007: 8; Paraskevas & Altinay, 2013: 158). While the concepts of pleasure, relaxation and security are embodied by tourism, crises create distress, fear, anxiety, trauma and panic (Santana, 2004: 300). According to a study conducted by the World Travel and Tourism Council, in order for tourism activities to become normal again; it takes 13 months after a terrorist attack, 21 months after an epidemic, 24 months after natural disasters, and 27 months in cases of political instability (cited in Zillmann, 2021). The travel and tourism industry is very sensitive to crises and is greatly affected by crises that create a negative perception of tourists (Kapiki, 2012: 20). In recent years, the global tourism industry has experienced many crises and disasters, including terrorist attacks, political instability, economic recession, biosecurity threats and natural disasters (Ritchie, 2004: 669). The consequences and effects of such crises go far beyond the immediate impact of the crisis, due to the interdependence of increasingly integrated global communications, tourism businesses, and transportation systems (Scott & Laws, 2006: 150). That is, global mass media tends to sensationalize negative events. This situation increases personal risk perception and serves to create an atmosphere of chaos (Pforr & Hosie, 2008: 249–250).

It can be stated that there are three components of the crisis: a triggering event that causes or has the potential to cause a significant change, the perception of not being able to cope with this change, and a threat to the existence of the foundation of the organization (Keown-McMullan, 1997: 4; Henderson, 2007: 3). Although the crises have become an integral part of commercial activities, it can be stated that one of the industries most affected by the crises is tourism. All crises have negative effects on tourist flows, but some events such as epidemics and terrorism have a great psychological negative impact on potential tourists not only during the crisis but also in the process following this period (Cavlek, 2002: 479–480). In fact, it is argued that tourism is an industry open to crises (Ghaderi, Som, & Wang, 2014: 638; Ertaş, Sel, Kırlar-Can, & Tütüncü, 2021: 1490). The sensitive nature of tourism to adverse events (Sigala, 2012: 1), the fact that there is always a crisis anywhere in the world, the certainty that another crisis will occur somewhere makes crises an almost permanent threat in terms of tourism (Pforr, 2006: 1). There is a conflict, contradiction and incompatibility between the tourism industry and the crises. On the one hand, tourism promotes visa-free travel and more comprehensive airline access, while on the other hand, it is advocated to increase security measures for crises (Tse, 2006: 38). Crises endanger the development of tourism not only with the damage they cause, but also with their unpredictability, revealing the sensitivity of tourism and how quickly its balance can be upset (Sausmarez, 2007: 701). As the tourism crisis is a complex issue of change and uncertainty, it poses significant challenges in terms of both supply and demand (Broshi-Chen & Mansfeld, 2021: 273). Crises occur at all levels of tourism operations, with varying degrees of severity, from environmental, economic and political disasters to accidents and sudden illness. At this point, the key factor is to keep the crisis under control and manage the situation without endangering tourists, no matter how dangerous the crisis is (Beeton, 2001: 422). Accepting a crisis and communicating effectively in the areas affected by the crisis will reduce the damage to image and reputation (Kash & Darling, 1998: 185). Crises require extraordinary practices rather than normal routine procedures, and each crisis has structurally different characteristics (Cushnahan, 2004: 327).

4.3 Terrorism

The concept of terrorism was used to describe the period from September 5, 1793 to July 27, 1794, when the French Revolution took violent and harsh measures against citizens suspected of being enemies of the Revolutionary Government. Afterwards, the resistance of the people against Napoleon's invasion of Spain was called guerra, meaning small war, and the word guerrilla is derived from this word (United Nations, 2018). According to Crenshaw (1981: 379), terrorism emerges in two ways: resistance to states and serving the interests of states. Whatever the reason for its emergence, terrorism conveys a violent message and the action goes beyond damaging the enemy's resources. The aim here is to attract the reaction of the masses. Primoratz (1990: 13) states that the purpose of terrorism is to oppress societies by creating fear. Kydd and Walter (2006: 50) argue that terrorism not only instills fear in target audiences, but also that governments and individuals support the case of terrorists. Terrorism has been defined by Laqueur (1996: 25) as the use of violence or threats with the aim of causing panic in a society, weakening or even overthrowing officials and bringing about political change. Pizam and Smith (2000: 123) defined terrorism as a systematic strategy that includes violence with the intention of intimidating the public or the region and creating fear for political, social and religious purposes such as assassination, hijacking, using explosives, sabotage and murder. Ruby (2002: 10–11) proposes three criteria that distinguish terrorism from other acts of violence. These criteria are as follows: terrorism is politically motivated, violence in terrorism is directed at civilians, and the violent policies of nation states are not counted as acts of terrorism. Terrorism is defined by Walters, Wallin, and Hartley (2019: 370) as a threatening act with a political, religious or ideological purpose that occurs in the form of harm or interference. Callaway and Harrelson-Stephens (2006: 681) state that terrorism is directed against civilians, unlike other forms of conflict where the targets are elements of the government.

Martin (2017: 2) classifies terrorism as new terrorism (the changing terror environment after 9/11), state terrorism (against counter-governments perceived as enemies), oppositional terrorism (ethnic or religious groups other than states), religious terrorism (tourism motivated by absolute faith), ideological terrorism, international, criminal opposition terrorism, gender-based terrorism. Purpura (2007: 17) states that there are various attempts to classify terrorism. Accordingly, one of these classifications is national and international terrorism. Another was cited as extreme leftism (USSR), far right (Ku Klux Klan), private or hate terrorism. Jenkins (2021), on the other hand, states that terrorism can be considered in three groups: revolutionary, sub-revolutionary and state-sponsored terrorism. Revolutionary terrorism aims to completely destroy a political system. Sub-revolutionary terrorism is aimed at changing the sociopolitical structure rather than the current regime. State-sponsored terrorism, on the other hand, includes acts of terrorism against the state's own citizens, factions within the government, other governments or groups. According to Ferreira, Graciano, Leal, and Costa (2019: 20) terrorism has changed form since the 1990s, this type of terrorism is called new tourism, terrorist groups are no longer isolated in their own territory, they operate in many nations, the resulting material losses and human losses are indicative of this change.

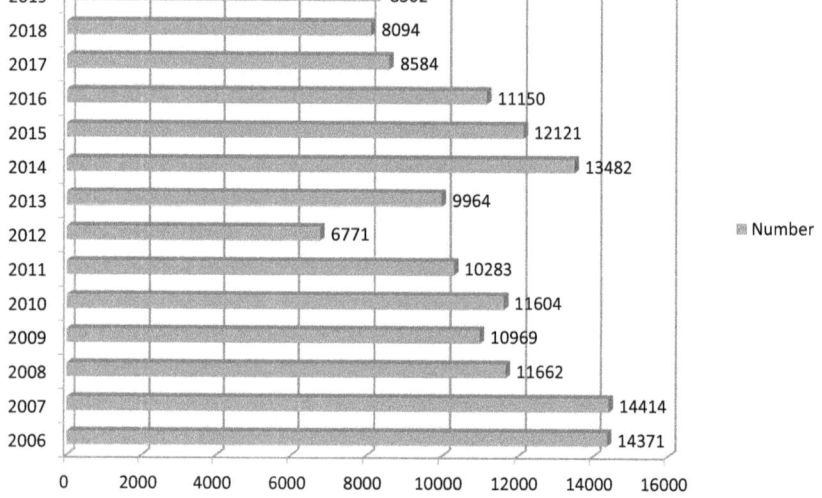

Graph. 2: Number of Terrorist Attacks Worldwide 2006–2019
Source: Statista, 2021

According to the 2019 report of the Global Terrorism Database, terrorist acts, which peaked in 2014, continued to decline for five consecutive years. While the Taliban was holding peace talks with the United States, Afghanistan experienced 21 % of all terrorist attacks worldwide in 2019, and 41 % of all people killed in terrorist attacks (including attackers) in 2019 died in Afghanistan. There were 13 attacks against tourists in 2018 and 12 attacks in 2019 (Global Terrorism Database, 2021). In 2019, 8302 terrorist attacks were recorded in the world. In 2006, the number of terrorist acts reached the lowest figure with 6771, but in 2014, the number of terrorist acts doubled with 13482 (Graph. 2). Most acts of terrorism occurred in Middle Eastern countries such as Afghanistan and Syria, with 1294 and 871 attacks, respectively (Statista, 2021).

4.3.1 Tourism and Terrorism

The killing of 11 Israeli athletes by Palestinian terrorists during the Munich Olympic Games in 1972 can be cited as the beginning of the relationship between tourism and terrorism. Approximately 800 million viewers watched the killing of athletes through television (Baker, 2014: 62). In fact, tourists are usually the special targets of terrorist organizations, because a terrorist act against tourists symbolizes an attack on the country where the tourists belong (Lepp & Gibson, 2003: 607). As a matter of fact, Aziz (1995: 92–94) states that the perception of Islam in the West and especially in the Western media should be examined in order to understand the terrorist acts of Islamic groups against tourists. In the Western mind, Islam's persistent connection with backwardness, oppression and terrorism has created Islam as a new enemy to replace the former Soviet bloc, and this has created Muslim terrorists. Another reason why terrorist groups target tourists is the prominence of tourism and tourists as another aspect of Western domination in various developing countries. In other words, tourists are the most clear and tangible representatives of wealthy and comfortable societies, challenging all moral, religious and social values of their society. Terrorist acts against the destination aim to scare and intimidate by making tourism an open target (Greenbaum & Hultquist, 2006: 113; Dragičević, Radić & Grbić, 2018: 64). While tourists can easily prefer safer destinations, local tourism industry and touristic destinations are exposed to profound effects after adverse events such as terrorist acts (Sönmez, Apostolopoulos, & Tarlow, 1999: 13). Tarlow (2014: 10) states that in terms of tourism, terrorist acts are related to promoting and sometimes gaining sympathy, and that many actions cause long-term damage to the image of the destination. Pizam and Fleischer (2002: 339) state that the factor that affects a destination the most is not the violence but the frequency of the terrorist act that occurs in the destination. In fact, the most negative effect of terrorist acts in terms of destinations is the decrease in demand for the destination. Avraham (2015: 226) states that every terrorist act reminds of the danger, that it may prevent tourist flows until this act is forgotten, but the incidence of terrorist acts in a particular region can completely end the tourism activity in that region. Drakos and Kutan (2003: 621) state that terrorist acts reduce foreign direct investments in destinations, cause additional advertising expenses to attract new or more tourists, cause reconstruction costs of damaged touristic facilities and cause new security expenditures to reduce terrorist threats.

Violent acts directly affect the image of a destination. It can be stated that individuals will not travel to places where they feel threatened. The negative reflections of this situation are the end of isolation, xenophobia and intercultural interaction (Tarlow, 2014: 1). Terrorism, a form of political instability and warfare, seeks to change policies and practices through intimidation and fear. The most common forms of terrorism include hostage-taking, murder, mass destruction of people and property (Timothy, 2006: 21), and unfortunately, terrorist attacks are difficult to prevent. It was previously believed that terrorists were psychopathic individuals who acted because of their mental instability. However, recent studies show that perpetrators are in a sense normal, aware of their actions and their consequences, and generally plan their actions very carefully and comprehensively (Teoman, 2017: 134). So why do terrorists target tourists? At this point, Albu (2016: 10) states that terrorists carry out attacks against tourists or destinations because of their desire to affect the economy of the region, attract international attention and promote themselves. Gustin (2021) states that killing and attempting to kill tourists is a brutal and indiscriminate terror tactic. Paraskevas and Arendell (2007: 1560) state that tourists and destinations are always easy targets for terrorists, that such actions are guaranteed to attract the attention of the international press, that they provide ease of conveying a political or ideological message, that the ability of the media to report on terrorist events, that the actions of tourists suggests that it will increase even more. Baker (2014: 61) states that terrorists target tourists in order to achieve ideological goals, punish citizens for supporting the government, and strengthen their political legitimacy claims by showing the government weak. Radić, Dragičević, and Sotošek (2018: 238) state that terrorists see tourists as easy targets and therefore their actions are aimed at tourists. According to Kurež and Prevolšek (2015), the reasons for tourists to be targets in terrorism are the opportunity to create many victims, the opportunity for publicity and media coverage, the opportunity for great economic damage and the destruction of national identity or symbols.

Major attacks against tourists are shown in the Tab. 2 (Paraskevas & Arendell, 2007: 1561–1562; Bysyuk, 2010: 7; Since 9/11, 2021; Albu, 2016: 11–14; BBC, 2021; Federation of American Scientist, 2021).

Tab. 2: Chronology of Terrorist Attacks against Tourists

\multicolumn{3}{l}{Chronology of Terrorist Attacks against Tourists}		
Date	Terrorist Attacks	Human Cost
1972	Hostage of Israeli Olympic Team in Munich, Germany	17 Death
1973	Attack and Hijacking at the Rome Airport	33 Death
1979	Grand Mosque Attack in Mecca	250 Death
1984	Restaurant Bombing in Spain	18 Death
1985	Attack on a Restaurant in El Salvador	13 Death
1985	Hijacking Italian Cruise Achille Lauro	1 Death
1985	Airport Attacks in Rome and Vienna	16 Death
1986	Aircraft Bombing in Greece	4 Death
1986	Bombing Nightclub in Berlin	2 Death
1986	Bombing Kimpo Airport in Seoul	5 Death
1988	Pan Am 103 Bombing	259 Death
1989	Bombing of UTA Flight 772	170 Death
1993	World Trade Center Bombing	6 Death
1995	Tokyo Subway Station Attack	12 Death
1995	Jerusalem Bus Attack	6 Death
1996	Manchester Truck Bombing	*
1996	Paris Subway Explosion	4 Death
1997	Sniper Attack on Empire State Building	2 Death
1997	Bombing Hotel Nacional in Havana	*
1997	Attack on foreign tourists in Luxor, Egypt	63 Death
2001	Hijacking in İstanbul	3 Death
2001	Bombing Nightclub in Tel-Aviv	*
2001	Bombing Restaurant in Jerusalem	15 Death
2001	Attack on World Trade Center	3025 Death
2002	Attack on Moscow Theatre	94 Death
2002	Attack on the Paradise Hotel Mombasa	15 Death
2002	Karachi Bus Bombing	13 Death
2002	Bali Bombing	202 Death
2003	Nightclub Bombing in Colombia	32 Death
2003	Suicide Bomb Attack in Morocco	43 Death
2003	Hotel Bombing in Indonesia	12 Death
2003	Rocket Attack on the al-Rashid Hotel in Baghdad	1 Death
2003	Bombing Neve Shalom and Beth Israel Synagogues in Istanbul	25 Death
2003	Suicide Truck Bombings in Istanbul	27 Death

(continued on next page)

Tab. 2: Continued

Chronology of Terrorist Attacks against Tourists		
2003	Suicide Bombing Moskow National Hotel	5 Death
2003	Bombing Nabil Restaurant in Baghdat	8 Death
2003	Bombing in Casablanca	33 Death
2003	Davao City Bombing Airport	21 Death
2003	Bombing Cnal Hotel in Baghdat	22 Death
2004	Train Bombing in Madrid	191 Death
2004	Two Russian Airlines Bombing	72 Death
2004	Bombing Superferry 14 in Manilla	63 Death
2004	Sinai Bombing	34 Death
2005	London Transport Bombing	52 Death
2005	Mumbai Railway Bombing	209 Death
2005	Suicide Bombing Sharia al-Moski in Cairo	3 Death
2005	Bombing Sharm el-Sheikh	88 Death
2005	Attack on a Tourist Bus in Kusadasi	5 Death
2005	Suicide Bombing in Bali	22 Death
2005	Attack on Hotel in Jordan	62 Death
2006	Dahab Bombing	23 Death
2006	Bombing Restaurant in Antalya and Kusadasi	3 Death
2006	Gun Attack on Amman	1 Death
2008	Siege of Mumbai	164 Death
2011	Breivik Shootings	77 Death
2013	Boston Marathon Bombing	5 Death
2013	Westgate Shopping Mall Bombings	67 Death
2015	Charlie Hebdo Shootings	12 Death
2015	Attack on Sousse Beach	38 Death
2015	Ankara Railway Station Bombing	103 Death
2015	Attack on Metrojet Flight 9268	217 Death
2015	Paris Attacks	138 Death
2016	Brussels Bombing	32 Death
2016	Orlando Gay Club Shooting	49 Death
2016	Attack on Nice, France	86 Death
2016	Attack on Berlin	12 Death
2017	Night Club Bombing in İstanbul	39 Death
2017	Bombing Manchester Arena	22 Death
2017	Barcelona and Cambrils Attack	16 Death

The impact of the September 11 attacks had a serious impact on international tourism. Despite the time passed, the tragedy of September 11 has changed international tourism and the mentality of international tourists through more comprehensive security checks at airports and tighter visa procedures. International arrivals in the USA decreased dramatically in 2001 and it took three years to reach the same level as 2000 (Liu & Pratt, 2017: 404). Essentially, the 9/11 actions are not directly aimed at tourism. However, in the immediate aftermath of this attack, activities and strategies that have not been implemented before have emerged in the control of international travel security at airports. The World Travel and Tourism Council (WTTC) states that unemployment in the tourism sector has increased worldwide after the event (Ferreira, Graciano, Leal, & Costa, 2019: 23). Goodrich (2002: 574) listed the security measures taken after the September 11 attacks as follows:

- ✓ 5000 new secret American National Guard members
- ✓ More private security guards at airports
- ✓ Only ticketed passengers can pass at the departure gate
- ✓ Detailed search of passengers at checkpoints
- ✓ Laptop computers are x-rayed
- ✓ Searching of flight personnel in more detail
- ✓ They have received scanners that can detect explosives.
- ✓ Participation of armed and civil security guards named Sky Marshal on some national and international flights.
- ✓ Studies on facial recognition technologies are in the form of
- ✓ More cameras recording daily activities
- ✓ Specially trained dogs that detect the location of bombs and
- ✓ Signing of border security agreement with Canada.

After September 11, there was a 0.6 % decrease in tourist arrivals worldwide. The continent of America (6 % decrease) was the continent most affected, followed by South Asia (4.5 %) and the Middle East (2.5 % decrease). A slight decrease of 0.7 % occurred in Europe. The Tab. 3 shows the comparison of international tourist arrivals in 2000 and 2001.

Tab. 3: International Tourist Arrivals between 2000 and 2001

Region	International Tourist Arrivals				
	(million)		Growth Rate (%)	Market Share (%)	
	2000	2001	2001/2000	2000	2001
World	696,8	692,6	–0,6	100	100
Africa	27,2	28,4	4,3	3,9	4,1
Americas	128,5	120,8	–6,0	18,4	17,4
East Aisa and the Pasific	109,2	115,2	5,5	15,7	16,6
Europe	402,5	399,7	–0,7	57,8	57,7
Middle East	23,2	22,7	–2,5	3,3	3,3
South Africa	6,1	5,8	–4,5	0,9	0,8

Source: World Tourism Organization, 2002

4.4 Epidemic Diseases

Another factor that negatively affects the choice of destination is epidemic diseases. The emergence of Covid-19 has integrated many concepts related to epidemics and pandemics into daily life. First of all, it is thought that it is useful to distinguish between these concepts. A pandemic is defined by Singer, Thompson, and Bonsall (2021: 1) as an epidemic that occurs worldwide or in a very large area, transcends international borders, and often affects large numbers of people. The World Health Organization, on the other hand, considered the pandemic as the worldwide spread of a new disease (WHO, 2021). In this definition emphasis is made on the particular that disease is new. This definition emphasizes the fact that the disease is new. However, the new emphasis on handling some diseases such as AIDS within the scope of a pandemic (Osman, 2008: 90) or after the emergence of shape-shifting versions of the 1918 H1N1 influenza virus (Girard et al., 2010: 4895) following its becoming a global epidemic (Girard et al., 2010: 4895) it can be stated that emphasis on being new does not affect whether a disease is pandemic or not. At this point the features of pandemic are specified as follows by Qiu et al. (2016–2017: 4–5):

✓ wide geographic extension
✓ disease mobility
✓ originality
✓ large amount of damage
✓ rapid spread
✓ minimal population immunity
✓ influence and contagion.

Endemics are constantly present in a given geographic area. A group of people can affect small areas or communities, such as all residents of a town or county, as well as larger areas such as countries or continents. The African origin of malaria is an example of an endemic situation (Grennan, 2021). Usually the amount of a particular disease present in a community is referred to as the baseline or endemic level of the disease (CDC, 2021). An epidemic is a disease that affects large numbers of people in a community, population, or region (Intermountain Healthcare, 2021). At this point, it can be stated that an epidemic (outbreak) can be called an epidemic if the dimensions of the geographical area covered are limited, and if the epidemic exceeds geographical borders, affects and threatens all humanity, it can be expressed as a pandemic. In other words, the main distinction is the geographical area where the epidemic affected.

4.4.1 Tourism and Epidemics

Epidemics have had a wide impact on human history throughout the ages. In fact, some of them changed the course of civilizations, caused destruction in the lands they affected, destroyed or transformed state structures, and sharpened political and social class distinctions (Buchillet, 2007: 514). The first and most important aspect of an epidemic is the loss of human life, the next is its effects on national or regional economies (European Parliament, 2020). Outbreaks do not have a clear starting point and can rise geographically very quickly (Boin, 2009: 368). The epidemics' elimination of mobility and the emergence of travel restrictions have greater repercussions, especially in countries whose economy depends on tourism industries. Yang, Zhang, and Chen (2020: 1) state that global epidemics greatly endanger the tourism industry as it depends on human mobility. In this case, a number of questions arise. When the epidemic is brought under control, will all the dangers disappear? Will tourism return to its normal course or will it completely change shape? Is there a possibility of similar epidemics that will affect tourism in the future? As a matter of fact, Leman (2021) stated that there were 28 different viruses that were trapped in a melting glacier 15000 years ago and had not been discovered before. Nordin (2005: 58) stated that climate change will have many negative effects on tourism, crises will occur more frequently due to global warming, and existing destinations may disappear. Yeoman and McMahon-Beattie (2006: 272) emphasize that the only way to make inferences about the future is to use the evidence in our environment and make a set of rational and creative assumptions that describe possible futures. This situation causes to question the consumerism and capitalist lens that contribute to mass growth in all tourism destinations; instead, it requires the creation of a system that promotes sustainable and equitable growth; it transforms experiences about what tourism is and what it should be (Benjamin, Dillette, & Alderman, 2020: 476). The Covid-19 pandemic is perhaps a step towards the collapse of globalization (Voth, 2020: 94).

Epidemics seriously affect tourist mobility, cause the spread of the disease internationally through modern aviation services, and create various restrictions on travel to some destinations (Pforr & Hoise, 2008: 253). Covid-19, the effects of which still continue, brought international travel to a standstill and once again revealed how vulnerable the industry is to crises (Farmaki, 2021: 1). The Covid-19 outbreak has been the biggest crisis experienced by the travel and tourism industry since the Second World War (Kaczmarek, Perez, Demir, & Zaremba, 2021: 1). Although various epidemic diseases such as SARS, Foot and Mouth Disease, Avian Flu have occurred before, none of them have had their effects on this scale. For example, 8096 people were infected between 2002 and 2003 due to SARS, and 774 people, that is, 9.6 % of those infected, died (Cooper, 2006: 119). Due to Covid-19, 185,291,530 people worldwide have been infected and 4,100,834 people have died (WHO, 2021). The travel and tourism industry has come to a standstill and countries have implemented various travel restrictions. The Graph. 3 shows the regional distribution of travel restrictions.

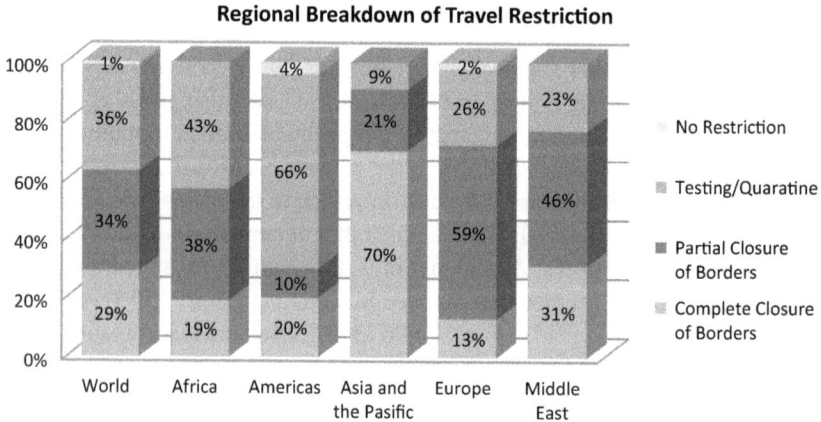

Graph. 3: Regional Breakdown of Travel Restriction
Source: (UNWTO, 2021)

According to the chart above, it is seen that only 1 % of the world has completely lifted travel restrictions. According to UNWTO's report dated 5 July 2021, 63 of 217 destinations, in other words 29 % of them are still completely closed. Thirty-four of these 63 destinations have closed their borders for at least 57 weeks. Seventy-nine destinations, that is, 34 % of all destinations, have partially closed their borders. Thirty-six percentage of the destinations, meaning. Seventy-eight destination countries, request a negative PCR test (Polymerase Chain Reaction) or antigen test upon entry as an international tourist. In some cases, quarantine is applied even if these tests are offered. The three countries that have completely lifted travel restrictions are Albania, Costa Rica and the Dominican Republic. The regional distribution of 63 destinations with completely closed borders is as follows: 32 destinations in Asia-Pacific, 10 destinations in Africa, 10 destinations in America, 7 destinations in Europe and 4 destinations in the Middle East. The distribution of destinations that have partially closed their borders is 32 destinations in Europe, 20 destinations in Africa, 10 destinations in Asia Pacific, 5 destinations in America, and 6 destinations in the Middle East. Of the destinations applying testing and/or quarantine, 34 are in the America, 23 are in Africa, 14 are in Europe, 4 are in Asia Pacific, and 3 are in the Middle East. Sixteen percentage of the world's destinations, that is 35 destinations, continue the mandatory quarantine application in addition to other restrictions regardless of where the passengers come from (UNWTO, 2021).

On July 1, 2021, the European Union implemented the Digital Covid certificate. This certificate is digital proof that the person has had the Covid-19 vaccine, the result of the PCR test is negative, or has recovered from the disease. The certificate has been put into practice in 27 European Union countries and 6 non-European Union countries (Iceland, Liechtenstein, Norway, San Marino, Switzerland, Vatican City) (European Commission, 2021). Although vaccination studies and practices to maintain social distance continue to control the epidemic, the emergence of various variants of the virus, the removal of restrictions on the borders of only three countries in the world, and the full border closure practices for countries where mutant virus has been seen in the last period are thought-provoking. The World Health Organization states that mutation of a virus has little or no effect on the virus's ability to cause disease or infection. However, depending on the genetic material of the virus, the mutated virus may spread more easily or in a more difficult way, the disease process may be more severe or milder (WHO, 2021). Although the Tourism Baramoter report published by UNWTO (2021) states that the speed of vaccination in some main destinations and policies to safely restart tourism, in particular the EU Digital Green Certificate, increase hopes for recovery in some markets. Abbas et al.

(2021: 1) state that the pandemic negatively affects the behavior and mental health of tourists, and states that because it seems impossible to prevent the transmission of the virus during travel, tourists cancel their holiday plans for fear of getting sick. Six hundred and fifty-five professional tourism workers were asked by Statista for their opinions on vaccination passports. Accordingly, 66 % of the employees stated that the vaccination passport is a good practice. However, the remaining part, considering the vaccination passport as a bad policy, being indecisive or not caring raises doubts about the success of this application (Graph. 4).

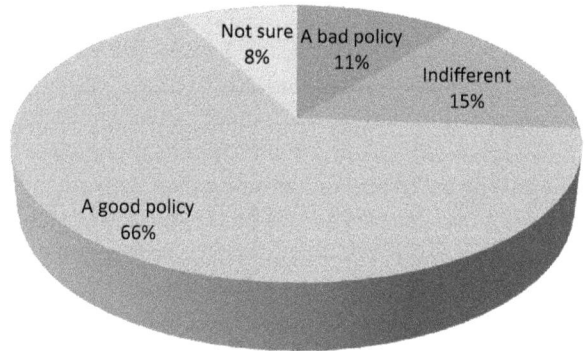

Graph. 4: Opinion on Vaccination among Business Travel Professionals
Source: (Statista, 2021)

The Covid-19 vaccine is not mandatory but encouraged by governments. However, the side effects of the vaccine are not clearly known. In addition, vaccination has become a necessity for individuals who want to travel internationally. Otherwise, it is either not possible to travel or mandatory quarantine is applied at the destinations. In addition, the application of the vaccination passport only between European countries may be a policy to develop tourism within the region. While a full framework has not yet been drawn about the effects of Covid-19 on people, are the measures taken by destinations sufficient? Even if the measures taken at the destinations are sufficient, will this be enough to attract tourists to the destination? The Graph. 5 shows the changes in international tourist arrivals in 2019, 2020 and 2021, comparatively.

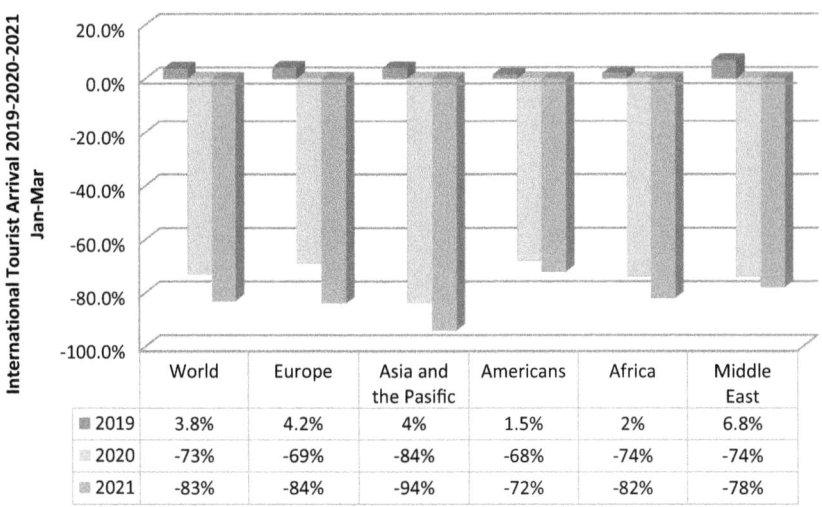

Graph. 5: International Tourist Arrivals 2019–2020–2021 Jan-Mar
Source: (UNWTO, 2021)

In the first three months of 2021, there was an 83 % decrease in tourist arrivals. Compared to 2020, the decline is higher due to travel restrictions and lower consumer confidence. The region with the highest decrease is the Asia Pacific destination. Other destinations are Europe 84 %, Africa 82 %, Middle East 78 % and America 72 %, from high to low, respectively. It is predicted that the industry will reach the figures for 2019 only in 2024 (UNWTO, 2021). The area most affected by the epidemic is the travel and tourism industry. Practices to control the pandemic have caused cancellations and postponements at an unprecedented level (Nicola et al., 2020: 188). Although the purpose of the measures taken was to protect human health in the first place, these measures dealt a heavy blow to international travel and the tourism industry (Çakır & Barakazi, 2020: 316). As a matter of fact, IATA stated in its annual report that the net loss in airlines in 2020 is 118.5 billion dollars (Graph. 6). While 38.9 million flights took place in 2019, the number of flights in 2020 is 16.4 million (IATA, 2020).

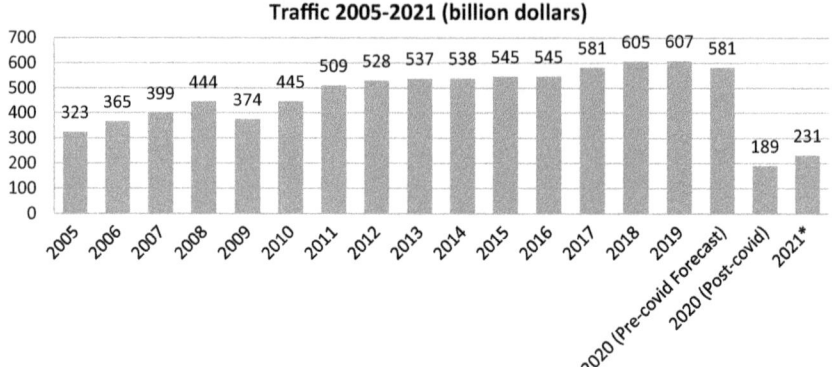

Graph. 6: Worldwide Revenue with Passengers in Air Traffic from 2005 to 2021
Source (Statista, 2021)

Conclusion

Tourism is very sensitive to crises because of its structural features, and emerging crises can give rise to long-term damages in destinations, and reflections of the crisis are felt more severely, mainly in countries whose economy is dependent on tourism. At this point, tourists' perception of the destination is highly important. Even though the qualities of the destination attract tourists, if tourists do not feel safe in the destination, this may cause changes and cancellations in holiday plans for the destination. In fact, the low perception of safety and security for a destination can give damage to the image and reputation of the destination, which causes irreparable losses.

Crises are unpredictable and they are high uncertainty events which can not be managed by means of classical methods. While the crises that occur in tourism sometimes have effect on a single facility or a single geographical area, the effect of some of them can be at a global level. The particular that there is a constant crisis in tourism creates an environment of chaos. But crises such as terrorist attacks or epidemics damage tourism as a whole, and threaten not only individuals or regions, but the whole world. In fact, the terrorist acts in the recent past and the pandemic we are nowadays experiencing have changed the lifestyles, habits and expectations of communities, made the tourism industry come to a standstill, perhaps reshaping the tourism phenomenon and putting it into a different form. This situation is an indicator of the fragility of not only tourism but also all humans. Tourism, which is one of the largest industries in the world, and every part of its industry are under threat and the future is quite uncertain. The precautions that are taken to prevent crises are insufficient with regards to both violent acts and epidemics. The fact that the world is pregnant to even bigger crises creates a frightening portrait. At this point, it is believed that a global cooperation is needed in order to minimize the impacts of the crises experienced.

Bibliography

Abbas, J., Mubeen, R., Iorember, P. T., Raza, S., and Mamirkulova, G. (2021). Exploring the Impact of COVID-19 on Tourism: Transformational Potential and Implications for a Sustainable Recovery of the Travel and Leisure Industry. *Current Research in Behavioral Sciences*, 2, 1–11.

Adler, S. (1977). Maslow's Need Hierarchy and the Adjustment of Immigrants. *The International Migration Review*, 11 (4), 444–451.

Adnan, M. (2014). Pakistan's Crisis Management: Examining Proactive and Reactive Strategies. *Journal of Political Studies*, 21 (1), 161–177.

Albu, C. E. (2016). Tourism and Terrorism: A Worldwide Perspective. *CES Working Papers*, 1–19.

Ali, Y., Shah, Z. A., and Khan, A. U. (2018). Post-Terrorism Image Recovery of Tourist Destination: A Qualitative Approach using Fuzzy-VIKOR. *Journal of Tourism Analysis*, 25 (2), 129–153.

Aliperti, G., Sandholz, S., Hagenlocher, M., Rizzi, F., Frey, M., and Garschagen, M. (2019). Tourism, Crisis and Disaster: An Interdisciplinary Approach. *Annals of Tourism Research*, 79, 1–5.

Alpaslan, C. M., and Mitroff, I. I. (2021). Exploring the Moral Foundations of Crisis Management. *Technological Forecasting & Social Change*, 167, 1–9.

Amerikanın Sesi. (2021, Temmuz 12). *Amerikanın Sesi*. Amerikanın Sesi: https://www.amerikaninsesi.com/a/venezuela-gecekondularinda-silahli-catismalar-suruyor/5959969.html (Access Date: 12.07.2021).

Amir, A. F., Ismail, M. N., and See, T. P. (2015). Sustainable Tourist Environment: Perception of International Women Travelers on Safety and Security in Kuala Lumpur. *Social and Behavioral Science*, 168, 123–133.

Anadolu Ajansı. (2021, Temmuz 12). *Anadolu Ajansı*. Anadolu Ajansı: https://www.aa.com.tr/tr/ulke-profilleri/honduras-cumhuriyeti/926405 (Access Date: 12.07.2021).

Augustine, N. R. (1995). Managing the Crisis You Tried to Prevent. In: *Harvard Business Review on Crisis Management*, Boston, Mass: Harvard Business School Publ. Corp., 147–161.

Avraham, E. (2015). Destination Image Repair during Crisis: Attracting Tourism during the Arab Spring Uprisings. *Tourism Management, 47*, 224–232.

Ayob, N. M., and Masron, T. (2014). Issues of Safety and Security: New Challenging to Malaysia Tourism Industry. *SHS Web of Conferences* (s. 1–10). Les Ulis: EDP Sciences.

Aziz, H. (1995). Understanding Attacks on Tourist in Egypt. *Tourism Management*, 16 (2), 91–95.

Baker, D. M. (2014). The Effects of Terrorism on the Travel and Tourism. *International Journal of Religious Tourism and Pilgrimage*, 58–67.

Baker, K., and Coulter, A. (2007). Terrorism and Tourism: The Vulnerability of Beach Vendors' Livelihoods in Bali. *Journal of Sustainable Tourism*, 2 (1), 249–266.

BBC. (2021, Temmuz 9). *BBC News World Aisa*. BBC News World Aisa: https://www.bbc.com/news/world-asia-19881138 (Access Date: 09.07.2021).

Beeton, S. (2001). Horseback Tourism in Victoria, Australia: Cooperative, Proactive Crisis Management. *Current Issues in Tourism*, 4 (5), 422–439.

Benjamin, S., Dillette, A., and Alderman, D. H. (2020). "We can't Return to Normal": Committing to Tourism Equity in the Post-pandemic Age. *Tourism Geographies*, 22 (3), 476–483.

Bhaduri, R. M. (2019). Leveraging Culture and Leadership in Crisis Management. *European Journal of Training and Development*, 43 (5/6) 534–549.

Blackman, D., and Ritchie, B. W. (2008). Tourism Crisis Management and Organizational Learning: The Role of Reflection in Developing Effective DMO Crisis Strategies. *Journal of Travel & Tourism Marketing*, 23 (2–4), 45–57.

Boin, A. (2009). The New World of Crises and Crisis Management: Implications for Policymaking and Research. *Review of Policy Research*, 26 (4), 367–377.

Boudreaux, B. (2005). Exploring a Multi-Stage Model of Crisis Management: Utilities, Hurricanes and Contingency. Florida: University of Florida.

Breda, Z., and Costa, C. (2006). Safety and Security Issues Affecting Inbound Tourism in the People's Republic of China. Y. Mansfeld, & A. Pizam içinde, *Tourism, Security and Safety From Theory to Practice* (s. 187–208). Burlington: Elsevier.

Brondoni, S. (2016). Global Tourism and Terrorism. Safety and Security Management. *Emercing Issues in Management*, (2) 7–16.

Broshi-Chen, O., and Mansfeld, Y. (2021). A Wasted Invitation to Innovate? Creativity and Innovation in Tourism Crisis Management: A QC&IM Approach. *Journal of Hospitality and Tourism Management*, 46, 272–283.

Buchillet, D. (2007). Epidemic Diseases in the Past: History, Philosophy and Religious Thought. In: M. Tibayrenc (Ed.), *Encyclopedia of Infectious Diseases: Modern Methodologies (Chapter 31)*, 517–524.

Buhalis, D. (2000). Marketing the Competitive Destination of the Future. *Tourism Management*, 21 (1), 97–116.

Bysyuk, V. (2010). *Impact of 9/11 Terrorist Attacks on US and International Tourism Development*. Vienna: Modul University.

Callaway, R., and Harrelson-Stephens, J. (2006). Toward a Theory of Terrorism: Human Security as a Determinant of Terrorism. *Studies in Conflict & Terrorism*, 29 (7), 679–702.

Cavlek, N. (2002). Tour Operators and Destination Safety. *Annals of Tourism Research*, 29 (2), 478–496.

CDC. (2021, Nisan 26). *Centers for Disease Control and Prevention*. Centers for Disease Control and Prevention: https://www.cdc.gov/csels/dsepd/ss1978/lesson1/section11.html (Access Date: 26.04.2021)

Coombs, W. T., and Laufe, D. (2018). Global Crisis Management – Current Research and Future Directions. *Journal of International Management*, 24 (3), 199–203.

Cooper, M. (2006). Japanese Tourism and the SARS Epidemic of 2003. *Journal of Travel & Tourism Marketing*, 19 (2–3), 117–131.

Crenshaw, M. (1981). The Causes of Terrorism. *Comparative Politics*, 13 (4), 379–399.

Cushnahan, G. (2004). Crisis Management in Small-Scale Tourism. *Journal of Travel & Tourism Marketing*, 15 (4), 323–338.

Çakır, P., and Barakazi, M. (2020). Koranavirüs Sürecinin Turizm Sektörüne Etkisi ve Salgına Karşı Alınan Tedbirler. *Anadolu Üniversitesi Sosyal Bilimler Dergisi*, 20 (3), 313–332.

Deutsche Well. (2021, Temmuz 12). *Deutsche Well*. Deutsche Well: https://www.dw.com/tr/yemende-sava%C5%9F%C4%B1n-alt%C4%B1nc%C4%B1-y%C4%B1l%C4%B1-%C3%A7%C3%B6z%C3%BCmden-uzak-%C3%B6l%C3%BCme-yak%C4%B1n/a-57016135 (Access Date: 12.07.2021)

Doepel, D. G. (1991). Crisis Management: The Psychological Dimension. *Industrial Crises Quarterly*, 5 (3), 177–188.

Dolnicar, S. (2005). Understanding Barriers to Leisure Travel: Tourist Fears as a Marketing Basis. *Journal of Vacation Marketing*, 11 (3), 197–208.

Dragičević, D., Radić, M. N., and Grbić, L. (2018). Terrorism as a Security Challange in Tourism Development. In *Proceedings Book of Tourism & Hospitality Industry*, 64–75.

Drakos, K., and Kutan, A. M. (2003). Regional Effects of Terrorism on Tourism in Three Mediterranean Countries. *Journal of Conflict Resolution*, 47 (5), 621–641.

Dyson, S. B., and Hart, P.. (2013). *Crisis Management*. Oxford: Oxford University Press.

Ertaş, M., Sel, Z. G., Kırlar-Can, B., and Tütüncü, Ö. (2021). Effects of Crisis on Crisis Management Practices: A Case from Turkish Tourism Enterprises. *Journal of Sustainable Tourism*, 29 (9),1490–1507.

European Commission. (2021, Temmuz 10). *EU Digital COVID Certificate*. EU Digital COVID Certificate: https://ec.europa.eu/info/live-work-travel-eu/coronavirus-response/safe-covid-19-vaccines-europeans/eu-digital-covid-certificate_en (Access Date: 10.07.2021).

European Parliament. (2020, Nisan 27). *Economic Impact of Epidemics and Pandemics*. European Parliament: https://www.europarl.europa.eu/RegData/etudes/BRIE/2020/646195/EPRS_BRI(2020)646195_EN.pdf (Access Date: 27.04.2021).

Evans, N., and Elphick, S. (2005). Models of Crisis Management: An Evaluation of their Value for Strategic Planning in the International Travel Industry. *International Journal of Tourism Research*, 7 (3), 135–150.

Farmaki, A. (2021). Memory and Forgetfulness in Tourism Crisis Research. *Tourism Management*, 83, 1–4.

Faulkner, B. (2001). Towards a Framework for Tourism Disaster Management. *Tourism Management*, 22 (2), 135–147.

Federation of American Scientist. (2021). *FAS*. FAS: https://fas.org/irp/threat/terror_chron.html (Access Date: 09.07.2021).

Ferreira, M. L., Graciano, P. F., Leal, S. R., and Costa, M. F. (2019). Night of Terror in the City of Light: Terrorist Acts in Paris and Brazilian Tourists' Assessment of Destination Image. *RBTUR*, 13 (1), 19–39.

Fuchs, G., and Reichel, A. (2011). An Exploratory Inquiry into Destination Risk Perceptions and Risk Reduction Strategies of First Time vs. Repeat Visitors to a Highly Volatile Destination. *Tourism Management*, 32 (2), 266–276.

Ghaderi, Z., Som, A. P., and Henderson, J. C. (2012). Tourism Crises and Island Destinations: Experiences in Penang, Malaysia. *Tourism Management Perspectives*, 2, 79–84.

Ghaderi, Z., Som, A. P., and Wang, J. (2014). Organizational Learning in Tourism Crisis Management: An Experience from Malaysia. *Journal of Travel & Tourism Marketing*, 31 (5), 627–648.

Girard, M. P., Tam, J. S., Assossou, O. M., and Kieny, M. P. (2010). The 2009 A (H1N1) Influenza Virus Pandemic: A Review. *Vaccine*, 28 (31), 4895–4902.

Global Terrorism Database. (2021). *Global Terrorism Overview: Terrorism in 2019*. Global Terrorism Database: https://www.start.umd.edu/pubs/START_GTD_GlobalTerrorismOverview2019_July2020.pdf (Access Date: 09.07.2021).

Goodrich, J. (2002). September 11, 2001 Attack on America: A Record of the Immediate Impacts and Reactions in the USA Travel and Tourism Industry. *Tourism Management*, 23 (6), 573–580.

Greenbaum, R. T., and Hultquist, A. (2006). The Economic Impact of Terrorist Incidents on the Italian Hospitality Industry. *Urban Affairs Review*, 42 (1), 113–130.

Grennan, D. (2021). *What is Pandemic*. Jama Network: https://jamanetwork.com/journals/jama/fullarticle/2726986 (Access Date: 26.04.2021).

Guillet, B. D., Lee, A., Law, R., and Leung, R. (2011). Factors Affecting Outbond Tourists' Destination Choice: The Case of Hong Kong. *Journal of Travel & Tourism Marketing*, 28 (5), 556–566.

Gustin, G. (2021). *Why do Terrorists Target Tourists?* Timeline: https://timeline.com/istanbul-why-a-decades-old-terrorist-tactic-has-gotten-uglier-and-more-random-e82a6bb5b030 (Access Date: 09.07.2021).

Hale, J. E., Dulek, R. E., and Hale, D. P. (2005). Crisis Response Communication Challanges. *Journal of Business Communication*, 42 (2), 112–134.

Hall, C. M. (2010). Crisis Events in Tourism: Subjects of Crisis in Tourism. *Current Issues in Tourism*, 13 (5), 401–417.

Hall, C. M., Timothy, D. J., and Duval, D. T. (2009). Security and Tourism: Towards a New Undestanding? In: C. M. Hall, D. J. Timothy, & D. T. Duval, *Safety and Security in Tourism: Relationship, Management and Marketing* (pp. 1–18). New York: Routledge.

Hannam, K. (2004). Tourism and Development II:Marketing Destinations, Experiences, and Crises. *Progress in Development Studies*, 4 (3), 256–263.

Henderson, J. C. (2007). *Tourism Crises: Causes, Consequences and Management*. Burlington: Elsevier.

Hernantes, J., Rich, E., Laugé, A., Labaka, L., and Sarriegi, J. M. (2013). Learning Before the Storm: Modeling Multiple Stakeholder Activities in Support of Crisis Management, a Practical Case. *Technological Forecasting & Social Change*, 80 (9), 1742–1755.

Hong, P., Huang, C., and Li, B. (2012). Crisis Management for SMEs: Insights from a Multiple-case Study. *International Journal of Business Excellence*, 5 (5), 535–553.

Howell, G., and Miller, R. (2006). How the Relationship between the Crisis Life Cycle and Mass Media Content can Better Inform Crisis Communication. *Prism Online*, 4 (1), 1–9.

IATA. (2020). *IATA Annual Review*. Amsterdam: IATA.

Intermountain Healthcare. (2021). *Intermountain Healthcare*. Intermountain Healthcare: https://intermountainhealthcare.org/blogs/topics/live-well/2020/04/whats-the-difference-between-a-pandemic-an-epidemic-endemic-and-an-outbreak/ (Access Date: 26.04.2021).

Jaques, T. (2007). Issue Management and Crisis Management: An Integrated, Non-linear, Relational Construct. *Public Relations Review*, 33 (2), 147–157.

Jia, Z., Shi, Y., Jia, Y., and Li, D. (2012). A Framework of Knowledge Management Systems for Tourism Crisis Management. *Procedia Engineering*, 29, 138–143.

Jing, Y., Deng, Y., and Zhao, Y. (2012). Study on Lifecycle-derived Law of Internet Public Opinion in Public Crisis Affairs. *International Conference on Management of e-Commerce and e-Government* (s. 178–182). Beijing: IEEE.

John Philip Jenkins. (2021). *Britannica*. https://www.britannica.com/topic/terrorism/Types-of-terrorism (Access Date: 09.07.2021).

Kaczmarek, T., Perez, K., Demir, E., and Zaremba, A. (2021). How to Survive a Pandemic: The Corporate Resiliency of Travel and Leisure Companies to the COVID-19 Outbreak. *Tourism Management*, 84, 1–11.

Kapiki, S. (2012). The Impact of Economic Crisis on Tourism and Hospitality. *Central European Review of Economics and Finance*, 2 (1), 19–30.

Karl, M., Muskat, B., and Ritchie, B. W. (2020). Which Travel Risks are more Salient for Destination Choice? An Examination of the Tourist's Decision-making Process. *Journal of Destination Marketing & Management*, 18, 1–11.

Kash, T. J., and Darling, J. R. (1998). Crisis Management: Prevention, Diagnosis and Intervention. *Leadership & Organizational Development Journal*, 19 (4), 179–186.

Keown-McMullan, C. (1997). Crisis: When does a Molehill Become a Mountain? *Disaster Prevention and Management: An International Journal*, 6 (1), 4–10.

Korstanje, M. E., and Tarlow, P. (2012). Being Lost: Tourism, Risk and Vulnerability in the Post-'9/11' Entertainment Industry. *Journal of Tourism and Cultural Change*, 10 (1), 22–33.

Kôvári, I., and Zimányi, K. (2011). Safety and Security in the Age of Global Tourism (The Changing Role and Conception of Safety and Security in Tourism). *Applied Studies in Agribusiness and Commerce*, 5 (3-4), 59-61.

Kozak, M., Crotts, J., and Law, R. (2007). The Impact of Perception of Risk on International Travellers. *International Journal of Tourism Research*, 9 (4), 233-242.

Kurež, B., and Prevolšek, B. (2015). Influence of Security Threats on Tourism Destination Development. *PREGLEDNI ČLANAK*, 9 (2), 159-168.

Kydd, A. H., and Walter, B. F. (2006). The Strategies of Terrorism. *International Security*, 31 (1), 49-80.

Laqueur, W. (1996). Postmodern Terrorism. *Foreign Affairs*, 75 (5), 24-36.

Laws, E., Prideaux, B., and Chon, K. (2007). Crisis Management in Tourism: Challenges for Managers and Researchers. In: E. Laws, B. Prideaux, & K. Chon, *Crisis Management in Tourism* (s. 1-12). Wallingford: CAB International.

Leman, J. (2021). *Popular Mechanism*. Popular Mechanism: https://www.popularmechanics.com/science/health/a30643717/viruses-found-melting-glacier/ (Access Date: 27.04.2021).

Lepp, A., and Gibson, H. (2003). Tourist Roles, Perceived Risk and International Tourism. *Annals of Tourism Research*, 30 (3), 606-624.

Lester, D., Hvezda, J., Sullivan, S., and Plaurde, R. (1983). Maslow's Hierarchy of Needs and Psychological Health. *The Journal of General Psychology*, 109 (1), 83-85.

Li, S., Blake, A., and Cooper, C. (2010). China's Tourism in a Global Financial Crisis: A Computable General Equilibrium Approach. *Current Issues in Tourism*, 13 (5), 435-453.

Liou, Y.-H. (2015). School Crisis Management: A Model of Dynamic Responsiveness to Crisis Life Cycle. *Educational Administration Quarterly*, 51 (2), 247-289.

Liu, A., and Pratt, S. (2017). Tourism's Vulnerability and Resilience to Terrorism. *Tourism Management*, 60, 404-417.

Liu, B., Pennington-Gray, L., and Krieger, J. (2016). Tourism Crisis Management: Can the Extended Parallel Process Model be Used to Understand Crisis Responses in the Cruise Industry? *Tourism Management*, 55, 310-321.

Loosemore, M. (1998). The Three Ironies of Crisis Management in Construction Projects. *International Journal of Project Management*, 16 (3), 139-144.

Maditinos, Z., and Vassiliadis, C. (2008). Crises and Disasters in Tourism Industry: Happen Locally – Affect Globally. *MIBES Conference 2008*

(s. 67–76). Larissa: Technological Institute of Larissa, School of Business and Economics.

Martens, H. M., Feldesz, K., and Merten, P. (2016). Crisis Management in Tourism – A Literature Based Approach on the Proactive Prediction of a Crisis and the Implementation of Prevention Measures. *Athens Journal of Tourism*, 3 (2), 89–102.

Martin, G. (2017). Types of Terrorism. M. Dawson, D. R. Kisku, P. Gupta, J. K. Sing, & W. Li içinde, *Developing Next Generation Countermeasures for Homeland Security Threat Prevention* (s. 1–16). Hershey: IGI Global.

Mathes, E. W. (1981). Maslow's Hierarchy of Needs as a Guide for Living. *Journal of Humanistic Psychology*, 21 (4), 69–72.

Mekinc, J., and Cvikl, H. (2013). The Structure of Security and Safety Crises in Tourism. *Journal of Tourism Services*, 4 (5/6), 38–50.

Micić, J., Denda, S., and Popescu, M. (2019). The Significance of the Risk-related Challanges in Tourist Destination Choice. *Journal of the Geographical Institute Jovan Cvijić*, 69 (1), 39–52.

Mitroff, I. I., Shrivastava, P., and Udwadia, F. E. (1987). Effective Crisis Management. *The Academy of Management Executive*, 1 (4), 283–292.

Mojtahedi, M., and Oo, B. L. (2017). Critical Attributes for Proactive Engagement of Stakeholders in Disaster Risk Management. *International Journal of Disaster Risk Reduction*, 21, 35–43.

Neumayer, E. (2004). The Impact of Political Violence on Tourism: Dynamic Econometric Estimation in a Cross-national Panel. *Journal of Conflict Resolution*, 48 (2), 259–281.

Nicola, M., Alsafi, Z., Sohrabi, C., Kerwan, A., Al-Jabir, A., Iosifidis, C., . . . Agha, R. (2020). The Socio-Economic Implications of the Coronavirus Pandemic (COVID-19): A Review. *Elsevier Public Health Emergency Collection*, 78, 185–193.

Nordin, S. (2005). *Tourism of Tomorrow: Travel Trends & Forces of Change*. Östersund: European Tourism Research Institute .

Norton, G. (1987). Tourism and International Terrorism. *Royal Institute of International Affairs*, 43 (2), 30–33.

Osman, A. S. (2008). HIV/AIDS in the Last 10 Years. *Eastern Mediterranean Health Journal*, 14, 90–96.

Paraskevas, A., and Altinay, L. (2013). Signal Detection as the First Line of Defence in Tourism Crisis Management. *Tourism Management*, 34, 158–171.

Paraskevas, A., and Arendell, B. (2007). A Strategic Framework for Terrorism Prevention and Mitigation in Tourism Destinations. *Tourism Management*, 28 (6), 1560–1573.

Pedersen, C. L., Ritter, T., and Benedetto, A. D. (2020). Managing through a Crisis: Managerial Implications for Business-to-business Firms. *Industrial Marketing Management*, 88, 314–322.

Penrose, J. M. (2000). The Role of Perception in Crisis Planning. *Public Relations Review*, 26 (2), 155–171.

Perles-Ribes, J. F., Ramón-Rodríguez, A. B., Rubia-Serrano, A., and Moreno-Izquierdo, L. (2016). Economic Crisis and Tourism Competitiveness in Spain: Permanent Effects or Transitory Shocks? *Current Issues in Tourism*, 19 (12), 1210–1234.

Pforr, C. (2006). *Tourism in Post-Crisis is Tourism in Pre-Crisis: A Review of the Literature on Crisis Management in Tourism*. Bentley: Curtin University of Technology School of Management Working Paper Series. Curtin University of Technology School of Management Working Paper Series: http://195.130.87.21:8080/dspace/bitstream/123456789/112/1/Tourism%20in%20post%20crisis%20is%20tourism%20in%20pre-crisis%20Pforr%20Christ.pdf (Access Date: 08.07.2021).

Pforr, C., and Hosie, P. J. (2008). Crisis Management in Tourism: Preparing for Recovery. *Journal of Travel & Tourism Marketing*, 23 (2–4), 249–264.

Pizam, A., and Fleischer, A. (2002). Severity versus Frequency of Acts of Terrorism: Which Has a Larger Impact on Tourism Demand? *Journal of Travel Research*, 40, 337–339

Pizam, A., and Smith, G. (2000). Tourism and Terrorism: A Quantitative Analysis of Major Terrorist Acts and Their Impact on Tourism Destinations. *Tourism Economics*, 6 (2), 123–138.

Primoratz, I. (1990). What Is Terrorism? *Journal of Applied Philosophy*, 7 (2), 129–138.

Purpura, P. P. (2007). *Terrorism and Homeland Security: An Introduction to Applications*. Burlington: Elsevier.

Qiu, W., Rutherford, S., Mao, A., and Chu, C. (2016–2017). The Pandemic and Its Impacts. *Health, Culture and Society*, 9, 1–11.

Racherla, P., and Hu, C. (2009). A Framework for Knowledge-Based Crisis Management in the Hospitality and Tourism Industry. *Cornell Hospitality Quarterly*, 50 (4), 561–577.

Radić, M. N., Dragičević, D., and Sotošek, M. B. (2018). The Tourism Led Terrorism Hypothesis: Evidence from Italy, Spain, UK, Germany and Turkey. *Journal of International Studies*, 11 (2), 236–249.

Ritchie, B. W. (2004). Chaos, Crises and Disasters: A Strategic Approach to Crisis Management in the Tourism Industry. *Tourism Management*, 25 (6), 669–683.

Rittichainuwat, B. N., and Chakraborty, G. (2009). Perceived Travel Risks Regarding Terrorism and Disease: The Case of Thailand. *Tourism Management*, 30 (3), 410–418.

Ruby, C. L. (2002). The Definition of Terrorism. *The Society for the Psychological Study of Social Issues*, 2 (1), 9–14.

Sahin, S., Ulubeyli, S., and Kazaza, A. (2015). Innovative Crisis Management in Construction: Approaches and the Process. *Social and Behavioral Sciences*, 195, 2298–2305.

Santana, G. (2004). Crisis Management and Tourism. *Journal of Travel & Tourism Marketing*, 15 (4), 299–321.

Sausmarez, N. d. (2007). Crisis Management, Tourism and Sustainability: The Role of Indicators. *Journal of Sustainable Tourism*, 15 (6), 700–714.

Scott, N., and Laws, E. (2006). Tourism Crises and Disasters: Enhancing Understanding of System Effects. *Journal of Travel & Tourism Marketing*, 19 (2–3), 149–158.

Seabra, C., Reis, P., and Abrantes, J. L. (2020). The influence of Terrorism in Tourism Arrivals: A Longitudinal Approach in a Mediterranean Country. *Annals of Tourism Research*, 80, 1–13.

Sigala, M. (2012). Social Media and Crisis Management in Tourism: Applications and Implications for Research. *Information Technology & Tourism*, 13 (4), 1–15.

Since 9/11. (2021). *Terrorism Timeline*. Since 9/11: https://since911.com/explore/terrorism-timeline (Access Date: 09.07.2021).

Singer, B., N. Thompson, R., and Bonsall, M. B. (2021). The Effect of the Definition of 'Pandemic' on Quantitative Assessments of Infectious Disease Outbreak Risk. *Scientifc Reports*, 11, 1–13.

Sirakaya, E., Sheppard, A. G., and McLellan, R. W. (1997). Assessment of the Relationship between Perceived Safety at a Vacation Site and Destination Choice Decisions: Extending teh Behavioral Decision-making Model. *The Council on Hotel, Restaurant and Institutional Education*, 21 (2), 1–10.

Sönmez, S. F., Apostolopoulos, Y., and Tarlow, P. (1999). Tourism in Crisis: Managing the Effects of Terrorism. *Journal of Travel Research*, 38, 13–18.

Sönmez, S., and Graefe, A. R. (1998). Determining Future Travel Behavior from Past Travel Experience and Perceptions of Risk and Safety. *Journal of Travel Research*, 37, 172–177.

Speakman, M., and Sharpley, R. (2012). A Chaos Theory Perspective on Destination Crisis Management: Evidence from Mexico. *Journal of Destination Marketing & Management*, 1 (1-2), 67–77.

Statista. (2021). *Statista*. Statista: https://www.statista.com/statistics/202864/number-of-terrorist-attacks-worldwide/ (Access Date: 09.07.2021).

Statista. (2021). *Statista*. Statista: https://www.statista.com/statistics/1229779/business-travelers-vaccination-passports-opinions-worldwide/ (Access Date: 10.07.2021).

Statista. (2021). *Statista*. Statista: https://www.statista.com/statistics/263042/worldwide-revenue-with-passengers-in-air-traffic/ (Access Date: 10.07.2021).

Stewart, M. C., and Wilson, B. G. (2016). The Dynamic Role of Social Media during Hurricane #Sandy: An Introduction of the STREMII Model to Weather the Storm of the Crisis Lifecycle. *Computers in Human Behavior*, 54, 639–646.

Tarlow, P. (2014). Tourism Oriented Policing and the Tourism Industry. *International Journal of Event Management Research*, 8 (1), 1–18.

Tarlow, P. (2014). *Tourism Security: Strategies for Effectively Managing Travel Risk and Safety*. Waltham: Elsevier.

Teoman, D. C. (2017). Terrorism and Tourism in Europe, New "Partners"? *European Journal of Geography*, 8 (2), 132–142.

Timothy, D. J. (2006). Safety and Security Issues in Tourism. In: D. Buhalis, & C. Costa, *Tourism Management and Dynamics: Trends, Management and Tools* (s. 19–27). Oxford: Elsevier.

Tokakis, V., Polychroniou, P., and Boustras, G. (2019). Crisis Management in Public Administration: The three Phases Model for Safety Incidents. *Safety Science*, 113, 37–43.

Tse, T. S. (2006). Crisis Management in Tourism. In: D. Buhalis, & C. Costa, *Tourism Management Dynamics: Trends, Management and Tools* (s. 28–38). Oxford: Elsevier.

United Nations (2018). *Introduction to International Terrorism*. Vienna: United Nations.

UNWTO. (2021). *Covid-19 Related Travel Restrictions a Global Review of Tourism*. Madrid: UNWTO.

UNWTO. (2021). *Tourism Barometer*. Tourism Barometer: https://www.e-unwto.org/doi/epdf/10.18111/wtobarometereng.2021.19.1.3 (Access Date: 10.07.2021).

UNWTO. (2021). *World Tourism Barometer*. Madrid: UNWTO.

Volo, S. (2008). Communicating Tourism Crises Through Destination Websites. *Journal of Travel & Tourism Marketing*, 23 (2–4), 83–93.

Voth, J. (2020). Trade and Travel in the Time of Epidemics. In: R. Baldwin, & B. W. Mauro, *Economics in the Time of Covid-19* (s. 93–96). London: CEPR.

Wahba, M. A., and Bridwell, L. G. (1976). Maslow Reconsidered: A Review of Research on the Need Hierarchy Theory. *Organizational Behavior and Human Performance*, 15 (2), 212–240.

Walters, G., Wallin, A., and Hartley, N. (2019). The Threat of Terrorism and Tourist Choice Behavior. *Journal of Travel Research*, 58 (3), 370–382.

WHO. (2021). *WHO*. WHO: https://www.who.int/csr/disease/swineflu/frequently_asked_questions/pandemic/en/ (Access Date: 26.04.2021).

WHO. (2021). *WHO*. WHO: https://www.who.int/news-room/feature-stories/detail/the-effects-of-virus-variants-on-covid-19-vaccines?gclid=CjwKCAjw55-HBhAHEiwARMCszipeIc-zGPpOYdvwsibtHErG3tbqWNkxSjC4ihqBPOpYeNOXcWFwTBoC_JIQAvD_BwE (Access Date: 10.07.2021).

WHO. (2021). *World Health Organization*. World Health Organization: https://www.who.int/emergencies/diseases/novel-coronavirus-2019 (Access Date: 09.07.2021).

World Tourism Organization. (2002). *Tourism Highlights*. Madrid: World Tourism Organzation.

Xiu-hua, Z., Zhi-yong, H., and Lu-qi, X. (2011). Tourism Destination Crisis Management Study: Based on the Crisis Life-cycle. *Conference on Information Systems for Crisis Response and Management* (s. 345–349). IEEE: Heilongjiang.

Yang, Y., Zhang, H., and Chen, X. (2020). Coronavirus Pandemic and Tourism: Dynamic Stochastic General Equilibrium Modeling of Infectious Disease Outbreak. *Annals of Tourism Research*, 83, 1–6.

Yeoman, I. (2011). *Tomorrow's Tourist: Scenarios & Trends*. Oxfordshire: Routledge.

Yeoman, I., and McMahon-Beattie, U. (2006). Tomorrow's Tourist and the Information Society. *Journal of Vacation Marketing*, 12 (3), 269–291.

Zillmann, C. (2021). *http://fortune.com/2015/11/30/terrorism-tourism-paris/* (Access Date: 13.04.2021).

Selcen Seda TURKSOY

Domestic Tourism post Covid-19 – An Opportunity to Revive Turkish Tourism?

A new type of coronavirus appeared in China and has spread worldwide, and its effects are still ongoing. When this new type of virus turned into an international public health emergency, it was named as 2019-nCoV or Covid-19 pandemic. The global economy was negatively affected, with losses of job and income. The closure in tourism destinations and attractions has led to the losses of jobs and a robust decline in economic benefits of tourism industry as well. As of 2020, the number of international travelers was a quarter of the previous year. Many countries have restricted international travel since the tourist flows were one of the ways of transmission of the virus. At this point, domestic tourism seems an important alternative for those who cannot go abroad due to international restrictions. It is also important for the recovery of tourism sector suffered from the pandemic seriously.

1 Introduction

Covid-19 first reported in China still continues to affect the world leading to social, economic problems and turndowns. Tourism has been one of the extremely and adversely affected sector by pandemic due to close interaction nature of tourism-related activities. International tourist arrivals have drastically declined. Travel restrictions imposed by countries one after another have reduced international mobility and deepened the negative impact of the pandemic on tourism. European Union (EU) countries, which closed their borders in March (2020) due to the Covid-19 pandemic, are constantly following a gradual approach to lift travel restrictions and update their list every two weeks (European Comission, 2021). Countries such as the USA, Canada, England, Japan, and Russia are silelim constantly renewing their travel restrictions as well. Despite the increase in the rate of vaccination and the vaccinated people, the spread of new variants prompts new travel restrictions. The slow pace in the fight against the pandemic makes it difficult to predict when travel will return into our lives. At this point, domestic tourism flows constitute an important opportunity for the recovery of countries and then the tourism sector. Indeed, the results of a recent regional survey in China indicate a constant preference for domestic trips

(McKinsey, 2020). Therefore, the study focuses on the current state and development of Turkish domestic tourism, and aims to provide suggestions for the sector which has contracted due to Covid-19 pandemic.

2 Pandemics and Tourism

The major pandemics up to date are over 20, and known as plague, cholera, influenza, AIDS and Covid-19 today. Black Death (black plague) was deadliest epidemics known. Three flu epidemics occurred with 10-year intervals in the 20th century; "Spanish Flu in 1918–19", "Asian Flu in 1957–58" and "Hong Kong Flu in 1968–70". The "Spanish Flu" (H1N1), the most severe pandemic in that era, caused 40–50 million people to die (LePan, 2020). Infectious diseases today still remain as a threat to humanity as the spread is easy and rapid through global trade and travels (Piret & Boivin, 2021). However, the World Health Organization (WHO) announced the new outbreak of the 21st century on December 31, 2019, with a series of "viral pneumonia" cases reported in China. The new "viral pneumonia" is an infectious disease caused by a new type of coronavirus called SARS-CoV-2. When this new type of virus turned into an international public health emergency, it was named the 2019-nCoV or Covid-19 pandemic.

Although crisis and disaster are terms that can be used interchangeably, they are apart in meaning. Moreira 2007 defined crisis as more predictable and observable, and disasters as negative events requiring immediate action, usually of shorter duration, caused by nature. Covid -19 is macro-level crisis affecting people across a wide area (Comcec, 2017). The tourism and travel industry encountered many crises through time such as terrorist attacks, coups, human rights violations, climate change, etc. These crises changed the direction of travels resulting in decrease in customer trust in the destination for a long time (Comcec, 2017). However today, it confronts a health-related crises affecting tourism worldwide- Covid 19.

Epidemics and pandemics have played an active role in social and economic change in history (Hall, Scott, & Gössling, 2020). For centuries to preserve the social structure, practices such as isolation, quarantine and border control have been used to prevent the spread of epidemics (Piret & Boivin, 2021). However, the literature on the interrelationship between pandemics and tourism and the long-term effects of pandemics on tourism is limited (Hall et al., 2020). SARS, H5N1 and H7N9 (Avian flu), and the Ebola outbreak in West Africa (2014–2016) were some of the epidemics impact the local population and tourists as well. During SARS outbreak, a short-term decline was observed in tourism demand and revenues until the epidemic was olarak değiştirelim over (Wang, 2009; Tang

& Wong, 2009). During Ebola outbreak, Hall et al., indicates that decline in demand is not limited to West Africa and the Congo, tourism in the whole continent was olarak değiştirelim suffered implying that sometimes silelim. suffered from olarak yazalım. the effects that are olarak değiştirelim not restricted to the image of the place associated with the pandemic, but is yerine are yazalım reflected to the entire region. The Covid-19 is somewhat different from the infectious diseases as the whole world encounters an unusual global health, social and economic emergency. The pandemic, which progressed slowly and was seen regionally in the beginning of 2020, then spread all over the world, causing the

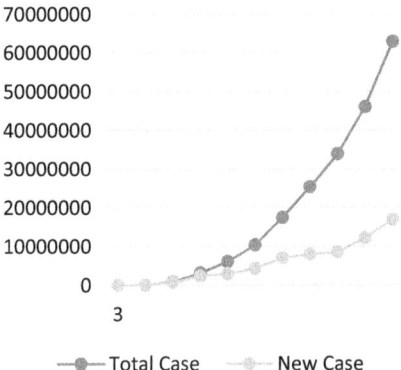

Fig. 1: Worldwide Covid-19 Monthly Cases (Total and New) (February-December 2020)
Source: WHO, 2021.

death of millions of people. Fig. 1 shows the number of cases in 2020 worldwide.

The pandemic began regionally in the first two months of 2020, spread rapidly and gradually throughout the world after March. While the number of cases increased, on the other hand the number of international arrivals showed a robust fall. The results of this pandemic, unlike the others, have been severe, and the sector has suffered a great deal.

3 The Effects of Covid-19 Pandemic on Tourism in Numbers

Since the emergence of the Covid-19 pandemic, global businesses have faced challenges. First, extensive quarantine measures were taken in several major cities of China and strict restriction measures were implemented, triggering the cessation of production, retail and trade activities within the country. Travels to and from China came to a standstill, with the spread of the virus worldwide many countries, South Korea and Italy in the first place, started to implement strict measures to control the epidemic in the following period. This situation adversely affected the companies involved in many line of business such as transportation, automotive, electronics, retail, automotive part supply, construction, textile, chemistry, agricultural food machinery equipment, metal, energy; and led to a decrease in their share values. With the quarantine measures, first local consumption expenditures and then international exports and imports have also decreased significantly especially in Europe and North America, and in the economies of other countries (Alianz, 2020).

The Covid-19 pandemic has an unprecedented impact on the hospitality industry (Gürsoy & Chi, 2020). In the first 10 months of 2020, international tourism flow decreased, that is; 900 million fewer foreign tourists traveled around the world compared to the same period of 2019 and generated export revenues decreased by $935 billion. This corresponds to a decrease of more than 10 times of the loss in 2009 (UNWTO, 2021b). Both international and domestic travel restrictions have caused the tourism industry to be more affected by the pandemic (Karabulut, Bilgin, Demir, & Doker, 2020). Depending on the spread rate of the pandemic around the world, the number of people participating in international tourism has decreased by 60 %–90 % since the first quarter of 2020. The total loss in the sector in 2020 has reached 75 % rate, and the number of international travelers has gone to 30 years ago. The months in which the highest

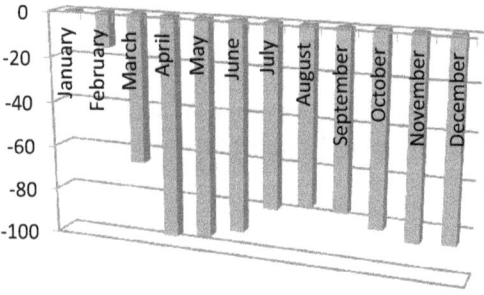

Fig. 2: International Tourist Arrivals by Month 2020 (Change %)
Source: UNWTO, 2021b.

decreases observed were April, May, June and November, respectively (Fig. 2).

When this worldwide decline (in January–November 2020) is examined on a continent basis, the biggest decline was seen in Asia Pacific with a decrease of 82 %, due to spread of the virus from China to the whole world. Middle East (−73 %), Africa (−69 %), America and Europe (−68 % each) followed Asia Pacific (UNWTO, 2020a). Based on the figures and the current trends, it is not easy for international tourism to recover. According to the World Tourism Organization the most important obstacle against global tourism movements is the ongoing travel restrictions. Although travel restrictions have been eased, some restrictive measures are still being implemented. Vaccination rate lagging behind the expectations and the newly emerging variants, economic problems, lack of coordination and low consumer confidence partially hinder the recovery of tourism. Although flights that restart with safe travel practices encourage tourism, it is expected that people will prefer to stay local and visit nearby destinations (leaning to domestic tourism) during the pandemic (OECD, 2020).

4 Government Supports to Tourism throughout Covid-19 Pandemic

Since the beginning of Covid-19 pandemic, travel and tourism industry has been negatively affected. With significant decrease in international and domestic travels countries have taken protective measures in order to prevent the probability of bankruptcy of tourism establishments and an increase in the number of unemployment in the industry. The volume of these measures is directly proportional to the economic power of the relevant country. When compared to other countries, a much larger support and bailout package was provided to the tourism industry in the USA. The supports generally constitute incentives for establishments or employees, but in some countries measures to improve domestic tourism are also put into practice (Tab. 1).

Tab. 1: Pandemic Support for Tourism Sector in Selected Countries* (2020)

Country	Support
Austria	Suspension of 2020 repayments of pre-pandemic loans.
Croatia	Postponing of tax payments. Liquidity support
Denmark	DKK 2 Billion loan guarantee support to airlines (SAS, etc.). Reservation cancellation support.
England	Liquidity support to be used in reimbursement for cancellations.
France olacak	Covering 70 % of gross salary and the whole of minimum wage. Pay up to € 10,000 to establishments with more than 50 % loss of revenue.
Germany	10-year loan support. € 9.000 / monthly up to 5 employees. € 15.000 / Monthly (3 month-support) up to 10 employees. € 3 Billion to Tour Operators (TUI, etc.)
Hungary	HUF 85 billion for the construction and renovation of a 100+ room facility in rural areas
Israel	ILS 300 Million support for re-openings.
Italy	Paying 80 % of salaries. € 600 for Seasonal Employees until March. € 1000 until August. Tax Reduction for establishments that comply with the distance rule. € 2 Billion for Accommodation Establishments. € 500 Million support for Airlines.
Luxembourg	€ 1250 support per employee (June-November 2020). Establishment of a tourism fund worth € 3 Million for non-profit tourism associations.
Portugal	Paying 2/3 of salaries. Grace period (2020) for leasing 4-year tourism loans. Support for tourism establishments affected by cancellations.
South Korea	Employment support up to 90 % of annual leave allowance for 6 months
Spain	Regulations to prevent job loss (until June 2020)
Switzerland	Liquidity support. Postponing loan repayments in 2020-2. Opportunity to extend loan term for facilities in mountainous, rural and border areas.
Turkey	Employment protective partial employment support. Liquidity support. Postponed principal and interests. Postponing hospitality tax. VAT discount on domestic flights (3 months)
USA	Support practices for employees. $25 Billion to passenger airlines; $3 Billion support for catering, luggage, ticketing, airplane cleaning and other industry contractors.

* Apart from the support presented in the table, countries may have direct or indirect support and incentives for the tourism industry under different names.
Source: OECD (2020), TURSAB (2020a).

Tab. 1 shows, employment protective incentives are the primary practice generally preferred by the countries. Partial or full salary supports for certain periods, reducing the tax burden on salaries (or exempting them from tax) are the most common measures. Other supports provided to establishments include postponing loan repayments until after 2020, extending the loan terms, liquidity support for establishments, and tax reductions. Some countries have chosen to support the airline companies that are flag carriers. Apart from these supports, Italy have chosen to give gratuitous traveler's check (500 €) for family to participate in domestic tourism (TURSAB, 2020a). Despite all these supports, it is predicted that it will take many years for international tourism to return before the Covid-19 pandemic.

5 Domestic Tourism to Recover the Tourism Industry in Turkey

5.1 The Importance of Tourism in Turkey

Turkey has great potential for tourism which is a constantly evolving industry. The number of 5.3 million visitors has exceeded 50 million since the nineties when the tourism industry began. It provides jobs for more than 1 million people (TUROFED, 2019). The strategy of tourism development based on the 3S principle has been abandoned. The Ministry of Tourism aims to silelim diversify tourism by integrating different types of tourism specific to the region with a sustainable approach (KTB, 2007).

Tab. 2: Economic Indicators for Tourism in Turkey (2011–2020)

Year	Tourist (Million)	Earnings ($Billion)	Expenditure per Tourist ($)	Total Earning/GDP
2011	36,15	28,12	778	3,4
2012	36,46	29,00	795	3,3
2013	39,23	32,31	824	3,4
2014	41,42	34,31	828	3,7
2015	41,62	31,46	756	3,7
2016	31,37	22,11	705	2,6
2017	38,62	26,29	681	3,1
2018	45,63	29,51	647	3,8
2019	51,86	34,52	666	4,6
2020	15,82	12,06	762	1,8

Source: TUIK (2020).

Turkey has taken the sixth place among the arrivals in top destinations pre-pandemic ranking (UNWTO, 2020a). While the number of visitors was 36 million in 2011, it exceeded 51 million in 2019 (Tab. 2). The only decrease in 10-year period is due to the loss of Russian tourism market as a political crisis has arisen between the 2 countries in 2016. Expanded bed capacity, increase in the number of direct scheduled and charter flights to the country and efficient promotion and marketing policies are among the reasons for the rise in tourism in the country. However, this upward trend was interrupted by the Pandemic. During the Covid-19 pandemic period, there was a 69 % decrease in the number of foreign tourists. The reasons for the decline are travel restrictions and strict quarantine measures applied in the country. The mutually loosened restrictions with Russia have led to maintain the number of tourists from that country.

In 2019, Turkey's tourism revenue was $34.5 billion (Tab. 2). The increase in revenue was by 1/4 compared to 2011. The increase rate in revenue fell behind the increase in the number of tourists. It can be explained through the devaluation of TL and the arrival of in low- and middle-income tourists. Tourism in the country, as a source of foreign exchange, is of great importance in financing and reducing current account deficit. The share of tourism in Turkey's GDP increased to 4.6 % in 2019 before outbreak of pandemic. It decreased to 1.7 % in 2020. The share of tourism revenues in total exports increased in the same period and reached 20 % (TUIK, 2019). The biggest spenders in Turkey were US, Japanese, Italian and Israeli tourists (TUIK, 2020). They spent more on food and beverage and transportation. Accommodation, clothing and souvenirs followed the two items.

According to the predictions of the World Tourism Organization, reaching 2019 levels in international tourism does not seem possible in the short term (UNWTO, 2021ab), and domestic tourism constitutes an important chance for the future of the sector, regardless of the situation of the Covid-19 pandemic. Based on estimations of the World Tourism Organization, the recovery of international tourism will begin in the fall of 2021 in parallel with the increase in consumer confidence and the decrease in restrictions (UNWTO, 2021b). It is quite obvious that prolongation of the pandemic will cause international arrivals to the country to remain limited; the source to keep the touristic facilities alive can only be obtained via the domestic market in Turkey.

5.2 Domestic Tourism in Turkey

Nine billion domestic trips were made worldwide in 2018 (UNWTO, 2020b). This figure is six times larger than international tourist arrivals; and some countries generate higher income from domestic tourism than active foreign tourism. While domestic tourism expenditures constitute 75 % of the total in OECD countries, domestic tourism expenditures in the EU are 1.8 times higher than inbound tourism expenditure. In terms of expenditure, the largest domestic tourism markets on a global scale are the USA with approximately $1 trillion, Germany with $249 billion, Japan with $201 billion and the United Kingdom with $154 billion (UNWTO, 2020a). A more balanced position between domestic and international tourism within the EU are observed. However, the added value created by foreign tourism arrivals in countries located in the Mediterranean basin, including Turkey, is considerably higher than the share of domestic tourism.

The income generated from active foreign tourism in our country is considerably higher than the share obtained from domestic tourism (TUIK, 2019). However, domestic tourism has played a major role in past crisis in Turkey. The past crises caused hoteliers in Turkey to lean to domestic tourists (Güzel, 2011). Gokdeniz and Dinc (2004) also indicate that domestic tourism has restarted and led to recovery phase in crisis in Turkish tourism mainly based on active foreign tourism. Cuhadar, Kervankiran, and Ongun (2020) state that sector representatives see domestic tourism as lifesaving particularly in times when international tourist numbers lag behind targets. In this respect, it is important to revive Turkish domestic tourism potential, during the covid-19 pandemic process. World Tourism Organization (UNWTO) also suggests that faster recovery of domestic tourism compared to international travel gives developed and developing countries the opportunity to get rid of the social and economic effects of the Covid-19 pandemic (UNWTO, 2020b).

Domestic tourism shows a slow growth and is not sufficiently encouraged by the government since it does not have a foreign exchange earning effect. However, it is actually a form of tourism with high social benefits that helps the participants to learn the natural and cultural values of their country (Kuşluvan, 2002). According to a study, domestic tourists in Turkey mostly visited the Aegean and Mediterranean coasts, big cities such as Istanbul and Ankara, and the cities of Trabzon, Bursa, Konya, Afyonkarahisar, Konya, Çanakkale and Adana between 2000 and 2015 and participated in tourism activities such as thermal, faith, highland, culture and skiing (Kervankıran & Çuhadar, 2017). In another study conducted during the pandemic process, the places with the highest domestic tourism activity in our country were Istanbul, East Marmara and West Aegean (Altuntas & Gok, 2021).

Tab. 3: Domestic Travel Indicators in Turkey (2011–2020)

Year	Domestic Travel (1000)	Overnight (1000)	Overnight per Travel
2011	65854	558270	8,5
2012	64922	556803	8,6
2013	68452	557459	8,1
2014	70894	575871	8,1
2015	71251	588786	8,3
2016	68450	605608	8,9
2017	77179	665194	8,6
2018	78523	633721	8,1
2019	78202	637070	8,2
2020	42847	469091	10,9

Source: TUIK (2017, 2019, 2020)

Travel, overnight stay and overnight stay per travel in Turkey in domestic tourism market are given in Tab. 3. The total number of domestic trips approached 80 million in 2019. The total number of overnight stays increased from 558 million to 637 million between 2011 and 2019. The average overnight stay fluctuated up to eight and nine per travel. 2016 was the year it reached the highest level with 8.9, when the number of tourists from Russia decreased significantly and terrorist attacks occurred. Overnight stay per travel reached a new milestone (10.9) in 2020- the year of pandemic, but the number of domestic trips fell substantially. The decrease in domestic trips throughout the country remained at 45 %. The number of overnight stays decreased 26 % compared to 2019. In other words, while the number of domestic trips decreased significantly during the pandemic period, the number of overnight stays of these people increased partially.

This situation shows that people are first hesitant about participating in touristic activities within the country when restrictions are lifted but when they participate, they increase their overnight stay per travel compared to previous periods. Average expenditure per trip in domestic tourism has been in the range of $100–$110 since 2017.

5.3 The SWOT Analysis of Domestic Tourism in Turkey

SWOT analysis guides the creation of development directions for countries, institutions and organizations, and businesses by identifying environmental relationships (Proctor, 1992). Glaister and Falshaw (1999) consider SWOT analysis to be one of the most respected and common tools used in strategic planning. In this aspect it is higly accepted technique in tourism industry as well. Narayan (2000) used SWOT analysis in his study, which aims to help the future growth policy decisions of the Fiji tourism. Miaoyan (2007) and Ramos, Salazar, and Gomes (2000) proposed a development model to develop tourism in China and Portugal. Bhatia (2013) analyzed the strengths, weaknesses, opportunities and threats of the Indian tourism industry to promote international tourism in country. In addition, the status and development potential of rural tourism and ecotourism in the Western Negev, Israel were evaluated using a different SWOT analysis (Kreiner & Wall, 2007). SWOT analysis was applied to identify target market groups, their strengths, weaknesses, threats and opportunities to guide the design of marketing strategies to promote domestic tourism in Rajasthan (Singh, 2013). In this aspect, in addition to facts and figures, SWOT analysis can assist environmental identification of the pre and post-pandemic state of Turkish domestic tourism and growth paths (Tab. 4).

Tab. 4: Strengths and Weaknesses, Opportunities and Threats of Domestic Tourism in Turkey before and after Covid-19 during the Pandemic Process

		After Covid-19		After Covid-19
Strengths	• Rich natural attractions • Cultural Heritage • Optimal Climate • Geographical location • Young workforce • Modern and well-maintained tourist facilities • Rich culinary culture	• Easy access to facilities and destinations (travel with their own vehicles) • Increased desire to go on a postponed vacation • Safe tourism practices	**Weaknesses** • Facility's appealing to middle and upper classes • High prices • Quality of employment • Seasonality • Spatial concentration	• Hygiene concerns • Covid-19 concerns • Lack of confidence in hygiene and certificates • Governmental restrictions within the country • Insufficient Covid-19 support packages
Opportunities	• Increasing interest in culture tourism • Developing transportation opportunities • Affordable vacation opportunities	• International travel restrictions • Covid-19 Pandemic • Vaccination • Increase in demand for small and boutique hotels • Increase in demand for caravan tourism • Increase in the number of installments in credit card payments. • Room price reduction in particular regions of the country • Increasing tourism-oriented domestic mobility • Increase in the number of travel blogs and itineraries • Distance education, remote/alternating working condition's enabling the mobility to increase	**Threats** • High taxes • Environmental problems • Cyber attacks • Regional investments exceeding the carrying capacity	• Reimplementation of the restrictions • loosened restrictions in tourist sending countries • Reservation cancellations • Bankruptcies in transportation sector • Online meeting and conferences • Slow progress of the fight against the pandemic • Constriction in global economy • Increase in demand for a limited number of destinations • Low consumer confidence

Source: TURSAB (2020a), Akgün, Turhan, Pamukcu and Turhan (2020), Altuntas and Gok (2021), Özdemir, Demirel and Demirel (2009), Güzel (2011)

The natural and cultural attractions, climate and geographical location, new facilities and rich culinary culture of Turkey has silelim have yapalım gained even more importance in pandemic conditions. As a result of the widely implementation of the ban and restriction in international travel, the desire to go on a holiday has directed people to domestic trips. Domestic tourism has been positively affected by the pandemic process with changes in travel patterns (preferred type of touristic product and the mode of transportation caravan tourism, traveling with their own vehicles etc.) Thus, the pandemic process has a positive impact on domestic tourism and has evolved the strengths of domestic tourism. Besides the chronic problems of domestic tourism such as seasonality, concentration in certain regions and high price, etc., pandemic and hygiene concerns and the insufficient support to the sector are the weaknesses that domestic tourism faces.

Declaration of new UNESCO cultural heritage sites and their promotion through films, etc., foundations of new museums, international travel restrictions, convenient routes for caravan tourism, competitive prices, increased number of travel programs and blogs in virtual and visual media offer new and large-scale opportunities for domestic tourism. On the other hand; high tax rates, environmental concerns and carrying capacity problems as well as the increase in the number of foreign tourists soon after loosening of international restrictions, consolidation and bankruptcies in transportation sector, reservation cancellations, global and national economic crises and increase in demand limited with a small number of destinations constitute the threats posed against domestic tourism in our country.

Conclusion and Suggestions

It is widely accepted that it will take years for international tourism arrivals to return to pre-pandemic times. Countries' acquirement of their target tourism revenue not only depends on their own policies, but also on international relations and the restrictions imposed by tourist sending countries towards their own citizens as long as the pandemic continues. At this point, the primarily probable solution lies in the encouragement of domestic tourism since Covid-19 pandemic brought international tourism flow to a standstill during the pandemic. In general the SWOT analysis shows that domestic tourism in Turkey has great potential in the period of Covid-19 pandemic. Growth paths and trends to develop domestic tourism during the pandemic period are listed below (Siteminder, 2021; TURSAB 2020ab; UNWTO, 2020b; UNWTO, 2021ab) and guide authorities and sector representatives.

Suggestions for Tourism Establishments
- As customers will be more sensitive about hygiene, the spread of safe tourism practices (safe facility/vehicle) should be supported.
- Distance is the other important factor for leisure. Businesses can target customers nearby.
- Information flow on the changes in consumer preferences and expectations during the pandemic should be updated.
- Dynamic practices can be adopted for free cancellation and cancellation fees.
- Pricing based on revenue management can be adopted.
- Policies and practices that shorten the check-in/out processes of groups can be developed (for example, via the app before they come to the facility).
- Alternative ways may be discussed for the use of idle meeting and conference centers.
- Instead of all-inclusive packages, half-board, bed and breakfast options may be offered.
- Hoteliers should pay attention to the increased hygiene and cleaning issues included in customer comments.
- More use of technology and robots can be benefitted during the pandemic.
- More efforts are to be made to increase visibility and clicks in the virtual world.
- Policies can be adopted to regain corporate customers.
- It should be taken into consideration that the number of blog followers on local restaurants, events and facilities may increase.
- Local events can be included more in marketing activities.
- Relaxing packages (spa/wellness) for customers with high anxiety levels can increase sales during the pandemic process.
- Delayed package tours can be updated and reorganized.

Suggestions for Authorities, Local Administrations and Unions
- Travel design developed for domestic tourism can be updated according to pandemic conditions.
- Applications (app) can be developed to share up-to-date information regarding restrictions.
- The concept of safe destination marketing can be focused on instead of marketing safe hotels.
- Customer-friendly practices can be adopted for requests such as cancellation and change during the pandemic process (international chain hotels announce their flexible policies to their customers on their websites).
- Positive news and blogs related to the region as well as the hotel can be brought to the forefront.
- International accreditation of health institutions in the destination can be encouraged.
- The number of installments can be increased in credit card payments. (TÜRSAB, 2020b). Low-interest vacation loan may be offered.
- Public service announcements and public relations activities can be carried out on the effectiveness of pandemic measures implemented by tourism establishments.
- Similar practices such as holiday support consumer loan, which was put into practice in the summer of 2020 (for salary customers, up to 10 thousand TL, 36 months with 6 months grace period and 0.67 % monthly interest rate) may become widespread (Turizmgunlugu, 2020).

As part of ensuring the expected growth in Turkey's domestic tourism, issues such as people's motivation to travel during Covid-19 and their expectations and preferences after restrictions, the effects of job and income losses on travel decisions should be examined first. The answers to these questions will be able to determine domestic tourism development direction. Besides these suggestions, it is important to create specific strategies for domestic tourism movements mainly concentrated in Istanbul, Marmara and Western Aegean to spread throughout the country in the pandemic period.

Bibliography

Akgün, T. Ö., Turhan, E. A., Pamukçu, H. and Turhan, A. (2020). *Covid-19 sürecinde Türkiye'de iç turizm talebi: Mevcut durum ve beklentiler*, Antalya.

Altuntas, F. and Gok, M. S. (2021). The effect of Covid-19 pandemic on domestic tourism: A DEMATEL method analysis on quarantine decisions, *International Journal of Hospitality Management*, 92, https://doi.org/10.1016/j.ijhm.2020.102719. (Access Date: December 7, 2021).

Bhatia, A. (2013). SWOT Analysis of Indian tourism Industry, *International Journal of Application or Innovation in Engineering & Management*, 2 (12), 44–49.

COMCEC (2017). *Risk and Crisis Management in Tourism Sector: Recovery from Crisis in the OIC Member Countries* (2021, 2 April). Ankara: Comcec Coordination Office.Web source: forbes.com/sites/tamarathiessen/2020/04/01/40- percent-less-flights-worlwide-air-travel-restrictions/#70aa335c6079 (Access Date: May 11, 2020).

Cuhadar, M., Kervankıran, I. and Ongun, U. (2020). Türkiye'de İç Turizm Hareketlerinin Tahmin Modellemesi: Karşılaştırmalı Bir Analiz. *Journal of Tourism and Gastronomy* Studies, 2 (8), 1113–1131.

European Comission (2021). Travel during the coronavirus pandemic (2021, 20 April). https://ec.europa.eu/info/live-work-travel-eu/coronavirus-response/travel-during-coronavirus-pandemic_en (Access Date: April 20, 2021).

Glaister, K. W. and Falshaw, J. R. (1999). Strategic planning still going strong, *Long Range Planning*, 32 (1), 107–16.

Gökdeniz, A. and ve Dinç, Y. (2004).Tur operatörlerinin bölgesel turizm pazarlarındaki yöre esnafına etkisi ve örnek bir araştırma, *Pazarlama Dünyası Dergisi*, Mart-Nisan, 9–21

Gursoy, D. and Chi, C. G. (2020). Effects of Covid-19 pandemic on hospitality industry: Review of the current situations and a research agenda, *Journal of Hospitality Marketing and Management*, 29 (5), 527–529.

Guzel, O. (2011). Türkiye'de iç turizm pazarı analizi ve pazarı canlandırmaya yönelik alternatif turizm olanakları, *Mustafa Kemal Üniversitesi Sosyal Bilimler Enstitüsü Dergisi*, 16 (8), 127–144.

Hall, C. M., Scott, D., and Gössling, S. (2020). Pandemics, transformations and tourism: Be careful what you wish for, *Tourism Geographies*, 22 (3), 577–598. https://doi.org/10.1080/14616688.2020.1759131 (Access Date: December 7, 2021).

Karabulut, G., Bilgin, M. H., Demir, E., and Doker, A. C. (2020). How pandemics affect tourism: International evidence, *Annals of Tourism Research*, 84, 1–5.

Kervankıran, İ. and Çuhadar, M. (2017). Türkiye'de İç Turizmin Gelişimi ve Mekânsal İstatistik Yöntemlerle Analizi, *Turizm Akademik Dergisi*, 4 (2), 1–18.

Kreiner, C. N. and Wall, G. (2007). Evaluating tourism potential: A SWOT analysis of the Western Negev, *Israel*, 55, 51–63.

KTB (2007). Türkiye Turizm Stratejisi Eylem Planı 2007–2013. https://yigm.ktb. gov.tr/Eklenti/906,ttstratejisi2023pdf.pdf?0 (Access Date: 2021, 27 July).

Kusluvan, Z. (2002). Türkiye'de İç Turizm Talebinin Analizi, *Journal of Tourism and Travel Resarch*, 2, 1–21.

Lepan, N. (2020). Visualizing the History of Pandemics, *Visualcapitalist*, https://www.visualcapitalist.com/history-of-pandemics-deadliest/ (Access Date: 2021, 14 March).

McKinsey (2020). The travel industry, turned upside down. https://www.mckinsey.com/~/media/mckinsey/industries/travel%20logistics%20and%20infrastructure/our%20insights/the%20travel%20industry%20turned%20upside%20down%20insights%20analysis%20and%20actions%20for%20travel%20executives/the-travel-industry-turned-upside-down-insights-analysis-and-actions-for-travel-executives.pdf (Access Date: 2021, 28 June).

Miaoyan, L. (2007). Preliminary study on development of industry tour in Liaoning province1/etude preliminmaire du developpement de l'industrie de tourisme dans la rovince du liaoning, *Canadian Social Science*, 3 (6), 36–9.

Narayan, P. K. (2000). Fiji's tourism industry: A SWOT analysis, *Journal of Tourism Studies*, 11 (2), 15–24.

OECD (2020). Rebuilding tourism for the future: Covid-19 policy responses and recovery. http://www.oecd.org/coronavirus/policy-responses/rebuilding-tourism-for-the-future-covid-19-policy-responses-and-recovery-bced9859/ (Access Date: 2021, 22 April).

Ozdemir, Y., Demirel, T., and Demirel N. Ç. (2009). Türkiye turizm sektörü için swot analizi ve strateji belirleme, Mersin Üniversitesi. *Turizm Isletmeciligi ve Otelcilik Yüksekokulu*, 21–24. Ekim 2009, Mersin.

Piret, J. and Boivin, G. (2021). Pandemics throughout History. In *Frontiers in Microbiology* (Vol. 11, p. 3594). https://www.frontiersin.org/article/10.3389/fmicb.2020.631736 (Access Date: 2021, 13 June).

Proctor, R. A. (1992). Structured and creative approaches to strategy formulation, *Management Research News*, 15 (1), 13–19.

Ramos, P., Salazar, A., and Gomes, J. (2000). Trends in Portuguese tourism: A content analysis of association and trade representative perspectives, *International Journal of Contemporary Hospitality Management*, 12 (7), 409–16.

Singh, R. (2013). Domestic Tourism in Rajasthan – Swot Analysis, *Pacific Business Review International*, 5 (11), 65–69.

Siteminder (2021). Coronavirus (Covid-19) recovery hope for hotels in the domestic tourism market. https://www.siteminder.com/r/marketing/hotel-digital-marketing/coronavirus-covid-19-recovery-hotels-domestic-tourism/ (Access Date: 2021, 1 April).

Tang, T. C. and Wong, K. N. (2009). The SARS epidemic and international visitor arrivals to Cambodia: Is the impactpermanent or Transitory? *Tourism Economics*, 15 (4), 883–890.

Turizm Gunlugu (2020). İşte 4 soruda tatil kredisinin detayları. https://www.turizmgunlugu.com/2020/06/01/coronavirus-tatil-destek-kredisi/ (Access Date: 2021, 12 April).

TUIK (2017). Hanehalkı Yurt İçi Turizm. https://data.tuik.gov.tr/Bulten/Index?p=Hanehalki-Yurt-Ici-Turizm-IV.Ceyrek:-Ekim-Aralik-ve-Yillik,-2017-27617 (Access Date: 2021, 15 June).

TUIK (2019). Hanehalkı Yurt İçi Turizm. https://data.tuik.gov.tr/Bulten/Index?p=Hanehalki-Yurt-Ici-Turizm-IV.Ceyrek:-Ekim-Aralik-ve-Yillik,-2019-33662 (Access Date: 2021, 15 June).

TUIK (2020). Hanehalkı Yurt İçi Turizm. https://tuikweb.tuik.gov.tr/PreHabe rBultenleri.do?id=33666 (Access Date: 2021, 15 June).

TUROFED (2019). Turizm Raporu, 2019/1. https://www.turofed.org.tr//panel/upload_system/pages_file/f0d8ac64de76e795a84a98a188b926b1.pdf (Access Date: 2021, 28 July).

TÜRSAB (2020a). *Türkiye ve Dünya Turizmi Değerlendirmesi*, Ankara.

TÜRSAB (2020b). Turizm harcamalarında kredi kartına taksit sınırı 18 aya çıkarıldı. https://www.tursab.org.tr/duyurular/turizm-harcamalarinda-kredi-kartina-taksit-siniri-18-aya-cikarildi (Access Date: 2021, 22 June).

UNWTO (2020a). *Interntational Tourism Highlights 2020*, Madrid.

UNWTO (2020b). UNWTO highlights potential of domestic tourism to help drive economic recovery in destinations worldwide. https://www.unwto.org/news/unwto-highlights-potential-of-domestic-tourism-to-help-drive-econo mic-recovery-in-destinations-worldwide#:~:text=article%20on%20linkedin-,UNWTO%20Highlights%20Potential%20of%20Domestic%20Tourism%20 to,Economic%20Recovery%20in%20Destinations%20Worldwide&text= As%20restrictions%20on%20travel%20begin,to%20explore%20their%20 own%20countries (Access Date: 2021, 15 June).

UNWTO (2021a). International tourism and covid-19. https://www.unwto.org/international-tourism-and-covid-19 (Access Date: 2021, 18 June).

UNWTO (2021b). Tourist arrivals down 87% in january 2021 as unwto calls for stronger coordination to restart tourism, *World Tourism Barometer*, Madrid.

Wang, Y. S. (2009). The impact of crisis events and macroeconomic activity on Taivan's international inbound tourism demand. *Tourism Management*, 30, 75–82.

WHO (2021). WHO Coronavirus (Covid-19) Dashboard. https://covid19.who.int/ (Access Date: 2021, 15 April).

İbrahim Tolga ÇOŞKUN

Economic Effects of Tourism and Tourism in the Period of Economic Crises

1 Introduction

There are different approaches in the classification of countries in which evaluations are made according to economic, political, strategic, socio-cultural or geographical features, and approaches in which economic criteria are dominant are widely used not only in academic studies but also in daily life. One of these classifications was made by the International Monetary Fund (IMF) and the United Nations (UN). Within the framework of this classification, the countries in the world are in three basic groups; It has been named as "Developed countries, Developing countries and underdeveloped countries". Regardless of which of these groups they are involved in, the geographical conditions of the countries significantly affect the economies of them. Countries tend to progress economically and develop compared to countries of a similar structure. Every country in the world, which has become a large global market, aims to take its share from this market in the direction of economic interests or to increase its current market share.

The most critical sectors leading the world economy are industry, service and agriculture. Industry is undoubtedly one of the most important sectors. In recent years, the service sector has increased its importance in the economic development of countries, rather than the industry and agriculture sectors. In increasing this importance, the service sector has the effect of being a faster and more revenue-generating sector than the industrial and agricultural sector. The tourism sector, which is one of the sub-sectors of the service sector, has the largest share in service exports in many countries. In addition, from the point of view of underdeveloped and developing countries, it is the most important factor that allows the entry of foreign money into these countries, which is also needed. Due to the rapid growth of the tourism sector in recent years and the fact that it is a sector that provides high income, developed countries also pay great attention to tourism and increase their tourism investments day by day, apart from industrial investments.

Tourism is a sector that is rapidly affected by economic crises, epidemics, social problems, geopolitical risks and uncertainties. For this reason, countries try to determine their short, medium and long-term plans and strategies correctly, and try to take quick and effective decisions in regional and global negative developments. Foreseeing negative developments and eliminating possible threats and uncertainties are the responsibility of not only country managers, but also industry representatives and academics. Today, risk and crisis management in the tourism industry has become a more important issue than in previous years. Accordingly, the correct handling and analysis of past economic crises, epidemics and negative developments will help minimize the risks to be encountered in the future. In this study, it is aimed to create a fundamental resource for managers, sector representatives and decision makers about global risks and crisis management, preparation for crises, the steps to be taken to eliminate uncertainties and risks, to reveal the economic effects of tourism and to explain the situation of tourism in the period of economic crises.

In the study, the economic importance of the tourism sector and the effects of the economic crises that have affected the whole world in the last 25 years on the tourism sector are discussed. The first part is the introduction, and in the second part, the economic importance of the tourism sector is explained according to the three main group classifications of the countries made by the International Monetary Fund (IMF) and the United Nations (UN). The third chapter is under the main title of "Global Economic Crises and Tourism" and consists of subtitles of 1997 Asian Economic Crisis, 1998 Russian Economic Crisis, 2007–2012 Global Economic Crisis and 2009 European Debt Crisis. In this section, the economic situation before and after the crisis, especially in the countries where these four crises started and heavily affected, is discussed, and the effects of the changes in tourism revenues and the number of tourists in the following years are explained. The results are presented in the last section.

2 Economic Importance of Tourism

In the classification created by the World Trade Organization, one of the 12 service sub-sectors of the service sector is tourism and travel-related services. Tourism and travel-related services are called as tourism sector in the most general sense. The tourism sector is the second largest economic sector in the world after industry. It is an important part of the economy for countries such as the USA, Japan, Germany, England, France and Turkey. The tourism sector, which has a volume of $ 2.2 trillion in Europe, contributes to the formation of an economy of $2.1 trillion in Asia. The sector, which has a volume of $2.75 trillion in 2018 and $2.84 trillion in 2019 in the global economy, is expected to reach a volume of $ 4 trillion by 2029. With the addition of factors such as investment, supply chain and income, the tourism sector, which reached a volume of $8.81 trillion in 2018 and $9.12 trillion in 2019, has become one of the most important elements of the global economy with the mobility it has created in the economies of countries. The sector, which is expected to contribute to the total employment in the world with a capacity of 420 million people until 2029, will also increase the investments by $1.45 trillion.

Tab. 1: International Tourism Receipts ($ Billion)

	2010	2015	2016	2017	2018
Total Tourism Export Receipts	1,152	1,445	1,469	1,590	1,712
International Tourism Receipts	0,980	1,228	1,254	1,352	1,458
International Passenger Transport	171	217	215	238	253

Source: The table was created using UNWTO reports containing the relevant years.

According to the data in the UNWTO (United Nations World Tourism Organization) World Tourism Barometer 2019, international tourism ranks third after fuel and chemicals in the 2018 world export rankings. In total export revenues in 2018, international tourism revenues reached $1.7 trillion, with $1.458 billion (visitor spending in destinations) and $253 billion in international passenger transportation revenue. It constitutes 29 % of world service exports and 7 % of total goods and services exports. These rates are quite high in some world regions. The data of the previous years of International Tourism Revenues, which are included in the statistical annex of the same barometer, are given in Tab. 1.

According to UNWTO World Tourism Barometer 2017 data, the tourism sector has grown approximately three times in the last 10 years and is expected to become the largest sector in the world in the next 20 years.

In underdeveloped or developing countries, tourism plays an important role in the context of insufficient savings, less export, and easier and faster acquisition of foreign exchange revenues necessary for national economies. Tourism, which is one of the labor-intensive sectors, has become a leading sector in many countries thanks to its positive effects such as realizing economic development, providing the needed foreign currency input, creating new income and employment opportunities. Other positive effects of tourism can be listed as income-generating effects, effects on balance of payments, effects on interregional development, employment-generating effects and effects on other sectors.

2.1 Economic Importance of Tourism for Developed Countries

The IMF's classification 39 countries are in the advanced economy group. Developed countries USA, UK, Japan, Italy, Germany, France, Canada, Netherlands, Portugal, Greece, Austria and Spain are among the top 20 countries visited by the most tourists in UNWTO World Tourism Barometer 2019. In the data of the same year, countries excluding Greece and the Netherlands are in the top 20 in tourism receipts. Greece, which is not in the top 20, is in the 21st place and the Netherlands is in the 24th place.

In developed economies, the share of international tourism in exports of goods and services was 6.8 % in 2010, 7.7 % in 2018 and 7.9 % in 2019. In the same years, the world average was 6.0 %, 6.7 % and 6.9 %, respectively. It is noteworthy that in developed economies, international tourism revenues are about 1 % above the world average and this share increases from year to year.

According to the data of the World Bank, the ratio of tourism receipts in exports in 2019 was 28.3 % in Greece, 23.6 % in Portugal, and 10.2 % in Austria. This rate is 9.2 % in the USA, which has the world's largest economy, and 8 % in France, which has the 6th largest economy. Based on these rates, the ratio of tourism receipts to export receipts has the largest share for some countries such as Greece and Portugal, which have relatively smaller economies among developed economies. The share of tourism in exports in countries with large economies such as France, Japan and Italy increased between 2008 and 2018. In Japan, which has a very important position in the export of technological and industrial products in the world, the share of tourism receipts in export receipts varied between 1 and 2 % between 1996 and 2012, while it was 2.4 % in 2014, 4.1 % in 2016 and It reached its historical peak of 5.4 % in 2019.

2.2 Economic Importance of Tourism for Developing Countries

Argentina, Russia, Brazil, China, India, Indonesia, Malaysia, Mexico, South Korea, South Africa, Taiwan, Thailand and Turkey are the main developing countries. According to 2019 data, all of these countries are among the 50 countries with the highest tourism income. China, Turkey, Mexico and Thailand are among the top 10 most visited countries. Malaysia ranks 14th and Russia 16th. While the average rate of tourism receipts of developing countries in export revenues in 2019 is 5.8 %, this rate is 20 % in Thailand, 17.2 % in Turkey, 9.3 % in Malaysia, 9.2 % in Indonesia, It is 8.6 % in South Africa and 7.1 % in Argentina. In addition, it has an almost continuously increasing rate in recent years. UNWTO Tourism Highlights 2015;

> *The market share of emerging economies increased from 30% in 1980 to 45% in 2014, and is expected to reach 57% by 2030, equivalent to over 1 billion international tourist arrivals. In 2030, 57% of international arrivals will be in emerging economy destinations (versus 30% in 1980) and 43% in advanced economy destinations (versus 70% in 1980).*

In 2019 data, this rate was 53.2 % in developed economies and 46.8 % in developing economies.

Tourism constitutes the most important export income of 83 % of developing countries, so the development of tourism is seen as an important goal for many countries, governments or regions. (Bahar & Bozkurt, 2010: 256). In many developing countries, tourism is regarded as an indispensable part of economic growth, foreign exchange earnings, employment growth and diversity, and an improvement strategy in the effective use of scarce resources (Sinclair, 1998: 1–51).

2.3 Economic Importance of Tourism for Underdeveloped Countries

In underdeveloped or developing countries, one of the main conditions for economic development is to increase foreign sales. Transition from traditional agricultural economy to industrial economy is a necessity for economic development and modernization in these countries. But such a change is possible with a large amount of capital, foreign currency gain, or external borrowing. This situation directs the country managers to the tourism sector in order to create the necessary financial resources for industrialization (İçöz & Kozak, 1998: 159).

Based on the Human Development Report of the United Nations Development Programme and published in 2020 using 2019 data, the number of underdeveloped countries is 33. In the export rates of 9 of these countries, tourism revenues are 20 % and above. The proportion of these countries and tourism receipts in export receipts is as follows; Gambia 48.3 %, Ethiopia 46.5 %, Eritrea 36.8 %, Haiti 34.9 %, Rwanda 28.3 %, Tanzania 27.2 %, Leberia 27.1 %, Madagascar 23.3 % and Sudan 20.9 %.

The economic benefits and impacts of tourism may be much greater in many underdeveloped countries, which are agricultural countries and are not common economies, than in developed countries' economies. Developing countries, especially after 1980, have started to give special importance to the development of the tourism sector due to reasons such as creating new employment areas, obtaining the required foreign exchange income and ensuring interregional economic balances. (Bahar & Kozak, 2006: 60).

3 Global Economic Crises and Tourism

In this section, financial developments that started in any country after 1997 and turned into a regional or global economic crisis and their effects on tourism will be discussed. In this direction, the Asian Economic crisis that started in Thailand in 1997 and affected many east Asian countries, the Russian Ruble crisis that started in Russia in 1998 and affected many neighboring countries, the 2007-2008 global economic crisis that started in the USA after 2007 and affected almost the entire world, and the 2009 European Debt Crisis, which started in Europe in 2009, triggered by global economic crisis, and affected many European countries, has been examined. The causes of the crisis in the countries where the crisis occured and heavily affected, and the general situations before and after the crisis, including tourism statistics, are discussed.

3.1 Asian Economic Crisis (1997-1999)

It is an economic crisis that started with the depreciation of the Thai currency Baht in 1997 and became a financial crisis in the Far East and Southeast Asian countries with the effect of the political crisis in the country, causing the international funds to change direction. China, Taiwan, Singapore and Vietnam are relatively less affected countries, while Japan is almost unaffected. The countries where the impact of the crisis is seen intensely are Thailand, Indonesia, South Korea, Hong Kong, Malaysia and the Philippines (Yamazawa, 1998: 332-333), (Goldstein, 1998: 1-6).

In this part, Thailand, the country where the crisis started and the effect of which was felt most deeply, is discussed. In the years of the crisis, Thailand was the most visited country in the South Asian group, but it was not an important point in the world. UNWTO Tourism market trends, 2000 edition; the financial and economic crisis in Asia caused arrivals to slow down immediately and, at times, bring them to a complete standstill. This slowdown in tourism has affected the already struggling economies even more deeply.

An important tourism attack was made during the crisis in Thailand, which was going through a difficult period economically. In the booklet "Co-operation and Partnerships in Tourism: A Global Perspective" prepared by the WTO Canadian Tourism Commission in 2003;

> *The 1998-1999 Amazing Thailand campaign coordinated by the Tourism Authority of Thailand (TAT) was a monumental effort to reposition Thailand as a unique destination for food, shopping cultural attractions and much more. Themed marketing campaigns were not new to Thailand. Indeed, 12 years before, Thailand had initiated the "Visit Thailand" campaign also to commemorate the king's birthday. The campaign was quite successful, but in the succeeding years tourism to Thailand had become stagnant. With the confluence of two high-profile events, the Asian Games and the King's Birthday Celebrations, the TAT thought that it was about time to have another big campaign. The objective was to attain a cumulative total of at least 17.8 million international visitors during the two-year period and boost Thailand's foreign exchange income to at least US $16 million. The goal for domestic tourism was to generate at least 122 million trips over the two-year campaign.*

Thailand received approximately 16.5 million international visitors over the two-year campaign period with continually increasing visitation thereafter.

The number of tourists and tourism receipts in the years after the crisis and the campaign period are shown in Fig. 1, and the progress in the world ranking according to the number of tourists and tourism receipts is shown in Fig. 2.

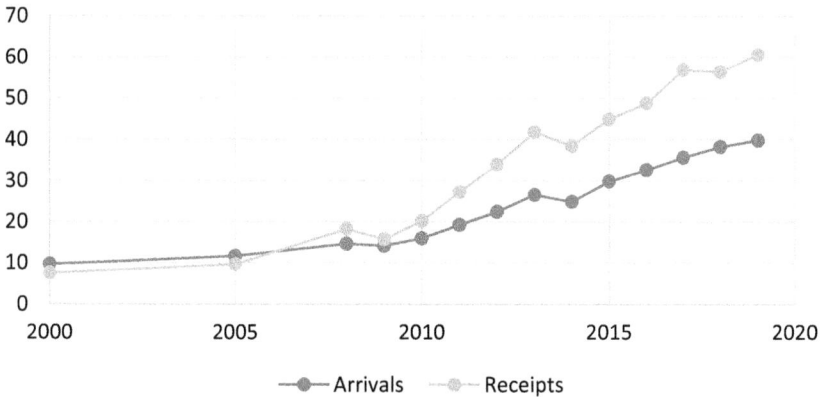

Fig. 1: Arrivals (Million People) and Receipts ($ Billion) in Post-crisis Years for Thailand

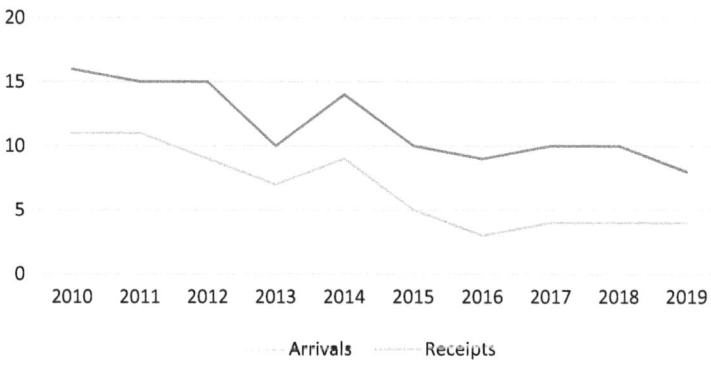

Fig. 2: Thailand's Progress in World Rankings

Although economic problems continued in Thailand until the end of 1999, it became a very important country in world tourism in the following years with the activities organized by TAT and the government's policies supporting tourism. While the ratio of tourism revenues to export revenues was 9.3 % in 2005, it was 10.4 % in 2010, 16.5 % in 2015 and 20.1 % in 2019, and this rate continues to increase.

3.2 Russian Economic Crisis (1998)

After the Asian crisis in 1997, oil prices decreased. Natural gas and oil account for more than half of Russia's export revenues. In addition, some Asian countries such as China and South Korea are among the countries to which Russia exports. While the barrel price of Brent oil was up to $45 at the end of 1996, it decreased below $20 in late 1998 with the effect of the Asian crisis. Oil exporting countries have been severely affected by this situation. Russia is among the top 10 countries in the world in terms of reserves. After the rapidly declining oil prices, the current account deficit problem occurred in Russia.

Apart from the current account deficit, another reason that led Russia to the crisis is high borrowing. The high rating of Russia in the years of the crisis facilitated the government's more short-term borrowing. High debts are also considered among the important causes of the crisis (Black et al., 1999: 1731–1808). As a result of the increased risk perception after the Asian crisis, the longer than expected Asian crisis, the support of new loans provided by the IMF, and Russia's insistence on the Ruble policy, a devaluation of over 30 % occurred.

Russia has an important position in world tourism. In 1999, it ranked 13th in the world in tourism expenditures, 11th in tourism revenues and 9th in the most visited destinations. According to UNWTO data, tourism receipts and expenditures between 1997 and 1999 are given in Tab. 2.

Tab. 2: Russia's International Tourism Receipts and Expenditures ($ Billion)

	1997	1998	1999	% Change 99/98
Receipts	7,1	6,5	7,8	19,4
Expenditure	10,1	8,3	7,4	-10,2

Source: The table was created using UNWTO reports containing the relevant years.

Tourism expenditures, which were $8,3 billion dollars in 1998, decreased by about 10,2 % and $7,4 billion in 1999. Although tourism receipts increased in the corresponding year, declining tourism expenditures had a negative impact on destination countries. In UNWTO Tourism Market Trends 2000, this situation is stated as follows;

> The start of economic recovery in major generating countries such as the Russian Federation, which in 1999 had not yet been achieved, caused the number of departures to neighbouring destinations in Central and Northern Europe to drop. In all the less developed countries in the region, factors linked with the allocation of resources for promotion and those which may have limited the operation of transport systems served to restrict tourism development in these destinations or to generate less arrivals from certain source markets. On the other hand, the Kosovo crisis and instability in the Russian market caused problems for mature destinations in Central and Eastern Europe such as Hungary (-14%), Poland (-4.4%) and the Czech Republic (-1.8%).

As in the case of Russia, even if there is no decrease in tourism receipts in the countries where the crisis began, there may be negativity in terms of destinations, and more countries may be economically, directly or indirectly affected by the domino effect of the crisis.

3.3 Global Economic Crisis (2007–2012)

In order to overcome the economic recession that followed the September 11, 2001 attacks in the United States until 2003, interest rates were reduced from 6.5 % to 3 %. As home loan costs also fell after the interest rate cut in 2003, the demand for housing increased rapidly. House prices also increased due to the increase in demand in the construction sector. Since there was an abundance of liquidity globally in the same period, loans were given even to individuals who could not afford to buy a house. With the increasing housing prices and the difficulties in payment, the loan repayments of individuals started to decrease and there was a boom in mortgaged housing. This financially unsustainable situation caused banks and financial institutions to declare huge losses and started the Mortgage crisis that broke out in 2007. Due to the expansionary monetary policy of the FED, instabilities were experienced in the financial markets after the intense fast and sloppy loans were given, the liquidity problems of the banks could not be resolved and the US stock market indices fell sharply due to the uncertainty environment. Because of its economic size and high trade volume, the financial problems of the USA have become a global crisis since 2008.

The most clear and obvious impact of the global crisis on tourism was experienced in 2009. International tourism revenues, which were $485 billion in 2001, increased steadily until 2008, reaching $989 billion. Even though the number of tourists fell during the 2003 SARS pandemic, tourism revenues increased by

$50 billion. Between 2004 and 2019, each year compared to the previous year international tourist arrivals (except 2008 and 2009) increased by 3.7 % and more. From 2007 to 2008, this increase was 1.9 %, and in 2009, when the crisis deepened its effects, it decreased by 4 %. There was a more dramatic decline in tourism receipts. Tourism receipts, which were 989 billion dollars in 2008, decreased by $88 billion in 2009 to $901 billion. In 2010, although the number of tourists was 952 million (929 million in 2008), revenues were $979 billion. In the following years, in parallel with the increasing number of tourists, tourism revenues did not increase at the same rate. As a result, it can be concluded that there is a differentiation in people's spending behavior for tourism even after the crisis period.

Noteworthy are the following statements in UNWTO Tourism Highlights 2010 on the global crisis;

> *The global economic recession aggravated by the uncertainty around the A(H1N1) influenza pandemic turned 2009 into one of the toughest years for the tourism sector. International tourist arrivals for business, leisure and other purposes totalled 880 million in 2009, corresponding to a worldwide decline of 4.2%. Growth returned in the last quarter of 2009, after 14 months of negative results.*

The 2008 economic crisis has the characteristic of being the most devastating crisis experienced since the 1929 world economic crisis. During the crisis, the real sector was hit the hardest, and both the employment contraction was experienced and the income declines caused by this contraction seriously affected the number of people participating in the international tourism movement. Fiscal and monetary measures must be implemented in harmony in order to get rid of the slowdown caused by the global crisis in the economies. It is important that the work to be carried out is very comprehensive, that there are projects covering sectors related to different sectors that provide employment and production opportunities. Public spending in sectors with a high multiplier effect helps to revive the economy in a short time. In this respect, it is seen that investments made in the tourism sector can be beneficial in times of crisis. In order not to be affected by the crisis, tourism enterprises should focus on innovative services, care about service quality and customer satisfaction, reduce costs and keep their cash flows strong, that is, they should become competitive (Yıldız & Durgun, 2010: 11–12).

3.4 European Debt Crisis 2009

The US financial crisis, the effects of which deepened in 2008, spread to the European Union countries in 2009 due to the intense trade relations between the European Union and the USA and the economic uncertainties of some countries in the Eurozone. Confidence in the Eurozone's single currency has made both individuals and countries more easily indebted. Some European countries, which are excluded from strong economies such as Germany, France and the United Kingdom, have lagged behind in development and competition, and their current account deficits have increased. Increasing budget deficits and public debt stocks have also increased financial fragility in European Union countries with relatively weaker economies.

Although the country where the crisis began in Europe is known as Iceland, Greece was the first country to request financial assistance from international financial institutions. The downgrade of the credit rating of institutions such as Fitch, S&P and Moody's has led to increased concerns that Greece will not be able to pay its debts, has caused the country's economy to struggle with problems that it has not been able to overcome since 2009 and to experience an economic recession that lasted for about 12 years. Similar to the Greek debt crisis was experienced in Ireland and Portugal. Spain, which is neighbor to Portugal, experienced a housing crisis, and the financial crisis was triggered by the increase in borrowing and the decrease in the economic growth rate. In Italy, which has the 8th largest economy in the world, the budget deficit decreased after 2009, but the public debt increased. Italy's economy contracted in 2013 and 2014 and entered recession. Imports also contracted due to a decrease in national income during the crisis. The countries most affected by the European debt crisis were Portugal, Ireland, Italy, Greece and Spain PIIGS (Portugal, Ireland, Italy, Greece, Spain).

As stated in the report published by UNWTO in December 2020, Spain ranks 2nd in 2018 and 2019 in tourism revenues, Italy 6th, Portugal 20th, Greece 22nd in 2018 and 21st in 2019, and Ireland 46th. Considering the free movement and commercial relations within the EU, it can be said that tourism has a great impact not only on the economies of these countries, but also on the economies of all European countries. In Tab. 3, there are tourism receipts, tourism expenditures, world rankings by years and countries of Portugal, Italy, Greece and Spain between 2008 and 2019.

Tab. 3: Tourism Receipts and Expenditures of some EU Countries for the Years 2008–2019

	Italy				Spain				Portugal				Greece			
	Expenditure		Receipts		Expenditure		Receipts		Expenditure		Receipts		Expenditure		Receipts	
	$	Rank	$	Rank	$	Rank	$	Rank	$	Rank	$	Rank	$	Rank	$	Rank
2008	30,8	6	45,7	4	20,3	12	61,6	2	4,3	41	10,9	27	3,9	43	17,1	12
2009	27,9	6	40,2	4	16,9	14	53,2	2	3,8	42	9,6	27	3,4	45	14,5	15
2010	27,1	8	38,8	5	16,8	17	52,5	2	3,9	42	10,1	27	2,9	49	12,7	21
2011	28,7	8	43	5	17,3	18	59,9	2	4,1	43	11,3	24	3,2	47	14,6	19
2012	26,4	10	41,2	6	15,4	20	58,2	2	3,8	45	11,1	27	2,3	*	13,4	22
2013	27	9	43,6	6	16,4	20	62,6	2	4,1	44	12,3	27	2,3	*	16,1	19
2014	28,8	8	45,5	7	18	20	61,5	2	4,4	44	13,8	27	2,7	*	17,8	19
2015	24,4	9	39,4	7	17,4	16	56,6	2	4	47	12,7	26	2,2	*	15,7	22
2016	25	9	40,2	7	19,3	14	60,5	2	4,3	46	14	25	2,2	*	14,6	23
2017	27,7	10	44,2	5	22,2	13	75,3	2	4,8	47	17,6	21	2,1	*	16,5	23
2018	30,1	10	49,3	6	26,8	11	81,7	2	5,4	46	20,1	20	#	*	19	22
2019	30,3	10	49,6	6	27,9	11	79,7	2	5,9	45	20,6	20	#	*	20,4	21

$ (Billion). * It is not among the top 50 countries in the ranking. # There is no information in the report of the World Tourism Organization.
Source: The table was created using UNWTO reports containing the relevant years.

Decreases or fluctuations in tourism expenditures in selected countries are noteworthy. Although there has been an increase in some years, the fluctuations have had a negative impact on the countries that receive tourists from these countries. In Greece, where the effects of the crisis continued, albeit partially, in 2021, tourism expenditures decreased significantly in the years after 2008.

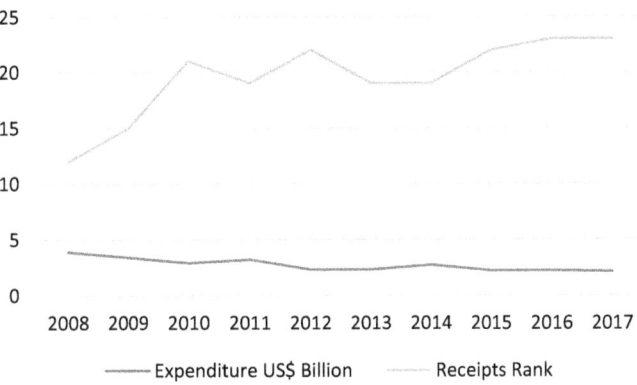

Fig. 3: World Ranking of Tourism Expenditures and Receipts of Greece in 2008–2017

According to the findings obtained from Tab. 3 and Fig. 3, it is noteworthy that Greece, which was most deeply affected by the 2009 European debt crisis, decreased continuously in tourism expenditures during and after the crisis, and in parallel, its tourism revenues declined in the world ranking.

Conclusion

Tourism is one of the indispensable sectors for all countries in the world. The economic importance of tourism is an undeniable fact. It is one of the sectors most rapidly affected by political and economic crises and epidemics. The tourism sector affects both the economies of countries where negative developments or crises are experienced and the economies of all other countries associated with those countries. The fact that it is a direct supporting power for many sectors and has indirect effects not only for the service sector but also for the industrial sector increases the importance of the tourism sector.

When the economic crisis after 1996–1997 are examined, it is clear that the Asian crisis, the US crisis, and the Russian and European crises have global effects. In this process, some countries were able to easily erase the effects of the crisis with alternative policies, while in some countries the negative effects continued for years. The examples that best explain this subject are; Thailand after the Asian crisis and Greece after the European debt crisis. Thailand has reached a very important point in world tourism after 2000 with its alternative and supportive tourism policies. Greece, which has very important advantages such as climate and geographical location, has not been able to erase the traces of the crisis it has experienced despite its high tourism potential, tourism investments have not been made and has lost its power in the tourism sector day by day. The Russian crisis and the subsequent economic losses in destinations affected by the Asian crisis are important for understanding the domino effect of the tourism economy.

Although the economies of developed countries have made progress thanks to industry and technology, they do not neglect their investments in the ever-growing tourism market. The fact that the share of international tourism revenues in exports of goods and services in many developed countries is high and these rates are increasing gradually is the most obvious indicator of this situation. Developing countries are in a similar situation and they are increasing their weight in the international tourism market with competitive policies. Even if less developed countries have industry, infrastructure or technological deficiencies, they should be able to offer options suitable for their geographical locations. They should be able to take initiatives for alternative tourism activities such as adventure, hunting, safari, river, gastronomy and cultural tourism.

Situations that are unfavorable with the right policies and initiatives can be turned into a favor. Each country should develop its own tourism policy, taking into account its geographical location and relationship with other countries. Political authorities, sector representatives and the academic community should work in coordination and solidarity on tourism. Short, medium and long-term plans should be made, and current and continuous studies should be carried out on crisis scenarios. It is necessary to be prepared for foreseeable situations, and crisis management should be given importance in the face of unforeseen or unexpected events. The regional and global effects of the economic crises experienced in the past should be carefully examined, and alternative tourism policies should be put into action immediately in case of possible risks and uncertainties. Governments also have important responsibilities in terms of tourism promotion policies, international organizations, and promotion of country potentials in other countries. Four seasons tourism opportunities should be created and investments should be increased. Apart from these, countries and sector representatives should take their own internal measures when necessary, and focus on competitive initiatives and marketing activities in alternative tourism types such as health, congress, faith, thermal, ecotourism, golf, balloon and cruise tourism.

Bibliography

Bahar, O. and Bozkurt, K. (2010). Gelişmekte Olan Ülkelerde Turizm-Ekonomik Büyüme İlişkisi: Dinamik Panel Veri Analizi. *Anatolia: Turizm Araştırmaları Dergisi,* 21(2), 255–265.

Bahar O. and Kozak M. (2006), *Turizm Ekonomisi,* Ankara: Detay Yayıncılık.

Black, B., Kraakman, R., and Tarassova, A. (1999). Russian Privitization and Corporate Governance: What Went Wrong. *Stanford Law Review,* 52 (6), 1731–1808.

Goldstein, M. (1998). *The Asian Financial Crisis.* Washington, DC: Institute for International Economics.

İçöz, O. and Kozak, M. (1998) *Turizm Ekonomisi.* Ankara: Turhan Kitapevi.

Sinclair, M. T. (1998). Tourism and Economic Development: A Survey. *The Journal of Development Studies,* 34(5), 1–51.

UNWTO (2011, 2012, 2014, 2015, 2017, 2018, 2019). *World Tourism Barometer.*

UNWTO (2015, 2019). *Tourism Highlights.*

UNWTO (2019). International *Tourism Highlights.*

Yamazawa, I. (1998). The Asian Economic Crisis and Japan. *The Developing Economies,* 36(3), 332–351.

Yıldız, Z. and Durgun, A. (2010). 2008 Küresel Ekonomik Krizi ve Turizm Sektörü Üzerine Etkileri. *Süleyman Demirel Üniversitesi Vizyoner Dergisi,* 2(1), 1–15.

Internet Sources

https://www.imf.org/en/Publications/WEO/Issues/2020/09/30/world-economic-outlook-october-2020

https://data.worldbank.org/indicator/ST.INT.RCPT.XP.ZS?most_recent_value_desc=true

http://hdr.undp.org/sites/default/files/hdr2020.pdf

https://www.aa.com.tr/tr/analiz/kovid-19-kuresel-turizm-sektorunu-ve-turkiyeyi-nasil-etkileyecek/1840883

https://www.e-unwto.org/doi/book/10.18111/9789284406012

https://www.macrotrends.net/1369/crude-oil-price-history-chart

Berna (KIRAN) BULĞURCU

Probing the Efficiency of Tourism Sector before and after Economic Crisis Periods: The Case of European Union Countries

This study aims to measure the tourism sector efficiency of 28 European countries with the data of three-year periods from 2008 to 2019 by using output-oriented and variable returns to scale Data Envelopment Analysis. The number of employees, tourism expenses, number of beds, and number of tourist arrivals were used as input indicators; the number of nights stayed and tourism revenue were used as output indicators. The long-term effects of the 2008 economic crisis on the tourism sector in the EU, which was most affected by the crisis, was evaluated separately by focusing on four three-year periods: 2008–2010, 2011–2013, 2014–2016 and 2017–2019. All reference sets for each period gave ideas to relatively inefficient European countries on efficiency improvement. According to the results of the analysis, recommendations are provided for decision-makers in tourism regarding making better decisions in the context of an economic crisis. Drawing on the results, it can be suggested that reaching highly effective outputs with improvement recommendations made countries strong while fighting a crisis. This research is thought to contribute to tourism sector literature at the macro-level evaluation.

1 Introduction

The effects of an economic crisis on the tourism have the potential to cause serious crises both within and outside the country in relations with other countries. One of these serious crises is the race to dominate the market. As a result of competition, the differences in tourism demand between countries cause a decrease in tourism revenues, which have a significant share in the economies of some countries. For this reason, tourism crises that may arise due to the differences in tourism demand are controllable crises that can be handled or even eliminated by regulations that are subject to competitiveness between countries. However, there are some crises that occur involuntarily and are unpredictable, such as occurring suddenly, showing uncertainty, affecting the image of organizations, profitability and even vital continuity (Demir, 2010: 15). According to Selamoğlu (2015: 27), the negative impact of the global economic crisis, which took place in late 2008 and had a strong impact as of 2010, on the European Union (EU) is beyond estimations. Many studies examining the effects of the global economic crisis for member countries with a common currency (Berberoğlu, 2011; Furceri & Mourougane, 2012; Yılmaz, 2013) have revealed how the Euro area experienced a great contraction in 2009. These unexpected contractions brought along the problem of not being able to control the national incomes, inflation, interest rates and changes in employment levels of the member states of the European Union, and the tourism sector also suffered from these problems. Indecision in the economic field has caused countries to change their tourism strategies and deteriorated the tourism supply and demand balance, just like the expense and income balance. For this reason, member countries have struggled to cope with the economic crisis in different ways within the same union in order to revive the tourism sector. Taking a country that manages the tourism sector well even in the crisis environment as a reference within the union is an advantage for the union countries that misalign their strategies in the tourism sector. To make better decisions about tourism activities, decision-making mechanisms generally use traditional decision-making techniques in an intuitive perspective. However, nowadays, decision makers began to use modern decision techniques which are based on performance measurement.

Performance measurement is a very broad concept. For this reason, it is frequently seen that measurement methods such as efficiency and effectiveness are used in performance measurement. In general terms, productivity refers to the relationship between the scarce resources (raw materials, people, capital, land, water, energy, etc.) used during production and the product obtained at the end of production, that is, the input-output ratio. Since the concept of efficiency is a fundamental problem that affects the living levels of all segments of society, it is a concept that should be emphasized and effectively managed by governments. It is hoped that the global economic problems that increase with the effects of globalization can be overcome with the right efficiency strategies. In fact, due to these problems, almost every country in the world has performance research and development centers and these centers make many measurement and evaluation models that are adapted within their own structure. If the performance measurement models are explained, these models are ratio analysis, which has a structure limited to a traditional input and output, parametric econometric models, and modern non-parametric techniques that we put in the category of new approaches. The most well-known modern non-parametric method is Data Envelopment Analysis (DEA). This approach evaluates the efficiency of homogeneous units producing the same output using the same input and compares each unit with the most efficient unit or units. In this respect, it seems more appropriate to use DEA, which is a homogeneous cluster understanding, in efficiency measurement compared to other approaches.

It is quite easy to calculate the efficiency ratio for the decision units in the case of a single input and a single output. However, mathematical operations by combining and formulating the relationship between multiple inputs and outputs are only possible with linear programming (Bakırcı, 2006: 119). In a problem, the desired goal in a linear programming model must be clear and measurable. It should be defined as a function and the degree of limitation of constraints that limit the degree of realization of this purpose should be known and expressed as linear equality or inequality. Therefore, it was decided to apply the DEA method within the scope of this study. DEA, which is a mathematics-based technique that calculates the relative efficiency of multiple decision-making units based on multiple inputs and outputs, was used in this study which was carried out in order to make an evaluation and comment on how EU member states can best transform the existing inputs into outputs in economic crisis environments (Liu et al., 2000: 145). In addition to being one of the most effective optimization methods in measuring relative efficiency, the fact that the calculation steps are not complicated is the reason for the selection of the method for the present study. The main purpose of the study is to evaluate the long-term effects of the 2008 economic crisis on the tourism sector in the EU, which was most affected by the crisis. In doing so, it is focused on the years between 2008 and 2019 by dividing this period into four three-year periods: 2008–2010, 2011–2013, 2014–2016 and 2017–2019.

One of the sub-objectives of the study is to explain how the DEA evaluates the findings, which will be used to analyse the data from the 28 EU member states, excluding Crotia since they joined the union in 2013, and identify the best and non-managers of this crisis process. According to Shurak (2001: 274), the global economy in the 21st century will be driven by information technologies, telecommunications and tourism. The importance of this study is also revealed by evaluating the existing power of the effects of tourism revenues on the national economies. After the introduction, in the second part, a literature review was carried out to determine the inputs and outputs that reveal the effectiveness of the tourism sector with the literature review of the period before and after the economic crisis. In the third part, after explaining what kind of method DEA is, the findings of the analysis carried out with the data collected in the fourth part are interpreted. The study, which is concluded with conclusions and recommendations, includes evaluations of the effects of economic crisis environments on the tourism sector.

2 Literature Review

The World Tourism Organization (WTO) defines crisis in the tourism sector as all the events that occur at unexpected times, reduce the confidence of tourists in the region and hinder the routine activities of tourism enterprises. The emerging globalized economic crises are also one of the main causes of the crisis in the tourism sector. In order to facilitate or eliminate all possible assumptions about the crisis, the effects and dimensions of the crisis, it is useful to examine in detail the studies on this subject in the literature.

In this respect, the studies carried out for the effectiveness of the tourism sector are important. Such studies are divided into two: analyses of micro units (hotels, travel departments, website of hotels, etc.) and macro level analyses. In this study, macro-level studies are included. Cracolici et al. (2008) used a 2001 economic data set to determine the economic efficiency of 103 Italian regions via two different models which are called stochastic and DEA frontier model. One output (bed nights) and three inputs (number of museums, monuments and archaeological sites, tourist school graduates divided by working age population and the labour units employed in the tourism sector) were used to evaluate tourism competitiveness of different regions. The ranking results of two models were compared and it showed that the efficiency coefficients from DEA frontier model was lower than the the stochastic frontier model coefficients.

Hadid et al. (2012) applied DEA to comparatively analyse the efficiency of tourism sector in 105 countries which were 34 developed and 71 developing countries. This study used two outputs such as the number of tourists and the expenditure per tourist in the country, while labour, rooms, natural resources and cultural resources were used as inputs. The differences between developed and developing countries were revealed in terms of tourism performance.

In Yi and Liang's (2014) research, they used DEA and Malmquist Index of the seven-year panel data for 21 cities in Guangdong Province to find the dynamics of tourism industry efficiency. This analysis ranked all cities by using the data of the input and output index values from a yearbook of Guangdong Tourism Statistics 2004 –2010.

Man and Zhang (2015) applied DEA to analyze the urban tourism industry in China. Both cities and regions were evaluated in terms of resources, technology and knowledge. The number of natural areas, the number of hotels and travel agencies were used as inputs, and the number of touristic trips and tourism income were used as outputs. Twenty-one provinces were compared according to their tourism efficiency according to the eastern, central and western regions.

In Atan and Arslanturk's (2015) research, the number of incoming passengers, number of mobility, tourism expenditures, passengers' tourism expenses for transportation needs and tourism expenditures for travel items were used as inputs while tourism revenues for all travel needs were used as outputs to evaluate tourism industry performance of World counries. The most referenced countries in the 5-year average are as follows: Seychelles (13 times), China (13 times), Sarajevo (12 times), Australia (8 times), Luxembourg (5 times), Panama (5 times) and Vanuatu (5 times).

Soysal-Kurt (2017) applied input-oriented DEA to evaluate tourism efficiency of 29 European countries for only 2013. While tourism expenses, number of employees and number of beds were used as inputs, tourism receipts, tourist arrivals and number of nights spent were used as outputs in the model. The aim of this research was to show how to realize the level of improvement offered from 16 efficient countries to 13 inefficient countries.

Bayrak and Rıza (2018) measured tourism potential of OECD countries by applying DEA. Number of arrivals, tourism expenditures and logistic performance index were inputs and also tourism revenues were output in this research. The USA, Australia, Spain, Luxembourg, Portugal, Turkey, New Zealand and Greece were determined as having high tourism potential countries.

In Chaabouni (2019)'s paper, tourism efficiency in China was tried to be measured on a regional level by applying DEA through two-stage double bootstrap approach. Bias corrected efficiency scores were helped to suggest many interesting and useful managerial inferences for tourism industry. By using indicators of the number of arrivals, tourism GDP, capital stock and labor, the level of efficiency and the status of economic development between three regions of China were revealed.

In the article of Barisic and Cvetkoska (2020), the effectiveness of travel and tourism impact for EU countries was analyzed using DEA. While domestic travel-tourism consumption and capital investment were used as input factors, the total contribution of travel and tourism to GDP and employment data were used as output factors and were expected to help policy makers to make the right decisions. The data was taken from only 2017. According to the results of the analysis, efficient and inefficient EU member countries in tourism were determined and a prospective research planning was decided to use data with longer time periods.

The motivation of this study is that the majority of DEA studies in the tourism sector are mainly carried out at the micro level. For this reason, by evaluating the tourism sector with DEA at the macro level, it was aimed to reveal how the countries should maintain the effectiveness of the tourism sector in times of economic crisis and what data they should take as a reference, with the output-oriented BCC DEA approach.

3 Methodology

3.1 Data Envelopment Analysis

Data Envelopment Analysis, which was developed by Charnes, Cooper and Rhodes in 1978 as a non-parametric method, is based on linear programming. This technique, whose mathematical steps are applied between homogeneous decision-making units (DMUs), calculates how efficiently the inputs are converted into outputs. The most characteristic features of DEA are as follows:

- It is based on the linear programming principle,
- Considering multiple inputs and outputs together,
- Inputs and outputs can have different units,
- It can be listed as comparing each decision unit only with the best decision unit.

Concepts such as effective limit and reference group in DEA are concepts that help to show efficiency in the graphical representation of DEA. It is the line segment that determines whether the effective boundary decision units are effective or not. Decision units that are not located on the active border are called ineffective decision units and their efficiency values are calculated depending on their distance from the active border. All the points of the efficient boundary that are enveloped by the upper boundary line are called the points that make up the production possibility set. The reference set, on the other hand, is the set of efficient decision units, which differ for each inefficient decision unit, are located on the efficient border and are shown as models for inefficient decision units. While the graphical representation of two and less than two input and output combinations can be possible on a plane, it is not possible to graphically represent the combinations that create more than three dimensions in the geometry space. Graphical representation of DEA is very difficult, since analysis of more than two decision variables is required in this study. The difference of DEA from other techniques that analyze efficiency is that the DEA method, which is used in many fields from the field of health to space studies, from the defense industry to the field of education, allows inputs and outputs to consist of different units and to compare each decision unit with only the best decision unit. In addition, since it is made according to the reference group formed by the best observations, instead of the mean function revealed by making statistical boundary estimation, the determined targets are formed by taking the best-performing units as an example, and this adds strength to the meaning and validity of the efficiency analyses made with DEA (Cooper et al, 2007: 25). The most basic models of DEA are CCR (Charnes-Cooper-Rhodes) and BCC (Banker-Charnes-Cooper) models. The analysis is facilitated by classifying these models into two main groups as constant returns to scale (CRS) and variable returns to scale (VRS). In addition, apart from the basic DEA models such as CCR and BCC models, there are additive model, slack-based activity measurement model and super-slack-based model among other DEA models for the measurement of efficiency that cannot be said to be input-oriented and output-oriented, that is, without focus. Being input-oriented means examining how the most appropriate amount of input should be used in order to obtain certain output amounts in the most efficient way and examining the changes that may occur in the number of inputs. Being output-oriented is the opposite of being input-oriented, and it is defined as examining the changes that may occur in output amounts by keeping the input amounts constant. In this study, the output-oriented BCC model with variable returns to scale was used. The most important reason for this is that the increase or decrease in the number of inputs that occur in economic crisis environments allows the output amounts to be seen and interpreted at different rates. Accordingly, the mathematical steps of the output-oriented BCC model will be explained. BCC is a new model introduced by Banker, Charnes, and Cooper (1984) with modifications on the CCR model. The only difference of the BCC model from the CCR model is that the sum of the values to be obtained as a result of the linear programming problem to be solved for each decision unit under the assumption of variable returns to scale is equal to one (Cooper et al., 2007: 91). Here, variable means the value that provides the information needed to construct an efficient possible input-output combination for an inefficient decision point. In short, the BCC model is a model that predicts technical efficiency and can distinguish between technical and scale efficiency. Understanding how activities are measured with fixed and variable returns to scale and what the concept of scale efficiency means is an important concept to understand before understanding the mathematical formulation of the BCC model. Fig. 1 explains CRS and VRS:

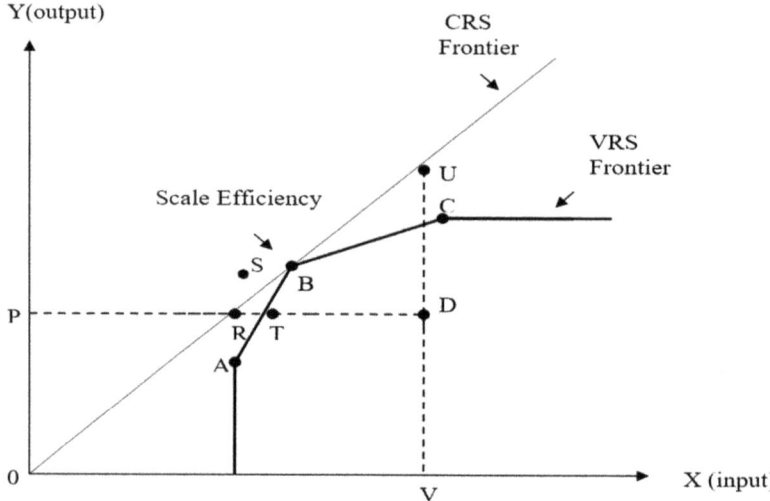

Fig. 1: Scale Efficiency
Source: Cooper et al., 2007: 338.

In Fig. 1, while A, B, C, D, R, S and T are decision units, it can be easily determined by using efficient frontiers whether each decision unit is effective or not. According to the figure, point D does not appear to be an efficient decision unit according to constant scale and variable scale. The reference group of D point is line segment AB in input-oriented models and line segment BC in output-oriented models. The technical efficiency values to be obtained with both approaches are different. Points A and C are not on the CRS efficient frontier and are not efficient according to the CRS assumption. However, since the points A and C are located on the efficient frontier according to the assumption of VRS, they are in the position of effective decision units. Scale efficiency or inefficiency is expressed in Fig. 1 as the distance between the efficient limit of fixed returns to scale and the efficient limit of variable returns to scale. Point B is the optimal point representing the scale efficiency, since it is located on both effective limits.

After understanding VRS, BCC mathematical formulation can be explained in the model as follows:

e_k = efficiency of kth DMU

u_{rk} = weight of rth output via kth DMU

v_{ik} = weight of ith output via kth DMU
Y_{rk} = Quantity of rth output of kth DMU
X_{ik} = Quantity of ith input of kth DMU
Y_{rj} = Quantity of rth output of jth DMU
X_{ij} = Quantity of ith input of jth DMU
N = Number of DMUs
s = Number of outputs
m = Number of inputs

The objective function aims to find the set of weights u and v that will maximize the efficiency of the decision unit. Constraints ensure that the weighted output/input ratio for each decision unit does not exceed 1. In this case, efficiency score will take a value between [0,1] (Kocakoç, 2003).

The formulation for this BCC model is presented in (Tone, 1996: 610):

Minimize $e_o = \sum_{i=1}^{m} v_i x_{io} - v_o$

Subject to = $\sum_{i=1}^{m} v_i x_{ij} - \sum_{r=1}^{s} u_r y_{rj} - v_o \geq 0$

$\sum_{r=1}^{s} u_r y_{ro} = 1$

$r = 1,......,s;\ i = 1,......,m$

Envelopment model of output-oriented BCC model is written as below:

Max $z_o = \emptyset$

Subject To= $\sum_{j}^{n} \lambda_j y_{rj} + Qy_o \leq 0$

$\sum_{j=1}^{n} \lambda_j x_{ij} + x_o \leq 0$

$\sum_{j=1}^{n} \lambda_j = 1$

$\lambda_j \geq 0;\ r = 1,......s;\ i = 1,......m;\ j = 1,......n$

Difference between BCC and CCR is about convexity constraint. In BCC model, sum of λs is equal to 1. These λs are calculated by using the steps of linear programming which will be solved for every DMU (Cooper et al., 2007: 89).

3.2 Data Collection

In this study, DEA is applied to 28 European Union countries which suffered from 2008 global economic crisis to evaluate tourism performance efficiency. The evaluation is performed by using variable returns to scale (VRS) and output-oriented BCC model via DEA Solver Pro software. In the analysis, four inputs and two outputs are determined as variables thought to be affective for the tourism performance efficiency of countries. All of the variables selected for the study are used commonly in the literature.

All data are collected from World Bank and OECD statistics. Due to limited data availability, Crotia is excluded from the analysis. Furthermore, although the United Kingdom is not a member of the European Union, it was a member during the economic crisis, and hence it is included in the analysis.

Tab. 1: Definition of the Variables

Inputs	Outputs
Number of employees (I1) (thousand) Tourism expenses (I2)(current US$) Number of beds (I3) Number of tourist arrivals (I4)	Number of nights stayed (O1) Tourism Revenue (O2) (current US$)

Source: Author

All used variables were decided with the tourism efficiency literature review. As seen in Tab. 1, the number of employees as the first input of these variables refers to the persons employed full-time and part-time in the tourism sector. Tourism expenses, the second input, are the money spent for all tourism activities by the government. Number of beds is related to how many beds tourist accommodation establishments have, as the third input. Number of tourist arrivals is the last input, and it means that the number of the tourist coming from abroad. According to the World Tourism Organization report (2016: 3), the number of arrivals shows the tourism size of countries.

Two important outputs of variables emerge when EU member countries, one of the unions most affected by the 2008 global crisis, use the existing inputs in order to increase the tourism efficiency in the crisis environment. The first of these is as follows: Number of nights stayed which means one night or over nights spent in tourist accommodation establishments. As a second output, tourism revenue describes the money spent by international tourists, and it shows the need of foreign capital.

The data of the study were organized in three-year periods as 2008–2010, the initial period of the economic crisis, the first period after the 2011–2013 crisis, the second period after the 2014–2016 crisis, and the third period after the 2017–2019 crisis. The arithmetic mean of 3-year data was used in the study, which was carried out in order to reveal how the activities of the EU member countries in the tourism sector were during and after the crisis.

Tab. 2: Descriptive Statistics

Year	Statistics	I1	I2	I3	I4	O1	O2
2008–2010	Max	38335	83556000000	5551669	42238595	212552531	62339333333
	Min	160	302333333	32248	786932	1531351	721333333
	Average	7854	13207226190	1024959	9677181	34195151	13146904762
	St. Dev.	9896	19485504721	1462823	12347754	50672278	16300977413
2011–2013	Max	39148	86913333333	5025677	48820873	245068646	66386333333
	Min	174	352000000	37015	922906	2223025	793333333
	Average	7744	13345785714	1055816	10800585	38813939	14089821429
	St. Dev.	9950	19654112464	1451823	14332696	58193941	17510272103
2014–2016	Max	40450	83558333333	5125614	56375861	276917709	67029000000
	Min	199	391000000	42656	1060568	2602340	915666667
	Average	7936	13316202381	1098260	13096232	49191171	14618071429
	St. Dev.	10267	19821964289	1499842	16959922	76309557	18267604195
2017–2019	Max	41993	92222000000	5108933	66244067	302007168	78979000000
	Min	238	506666667	45989	1034911	2550955	1008000000
	Average	8256	14916178571	1132104	15183311	54937179	16440083333
	St. Dev.	10630	21412138440	1526263	19047925	81538772	20213786764

Source: Author

The descriptive statistics presented in Tab. 2 are data on the three-year averages of input and output variables mentioned in Tab. 1 for 28 EU countries. According to the standard deviation values, the data show a wide range of distribution.

4 Probing the Efficiency of Tourism Sector before and after Economic Crisis Periods

The global economic crisis that began in the USA in 2008 gradually spread to other countries and took over EU member countries as well. Particularly among the EU countries which had weak economies and were struck hard by the crisis, the questions were raised about the vulnerability and fragility of the EU as a community. Tourism revenues fell into a decline after the crisis (European Commission, 2011: 120). The unfavorable circumstances that may be caused by the crisis in other areas of the Mediterranean region have been considered to provide solutions regarding how tourism revenues can be boosted. The output-oriented BCC DEA model is used to determine the EU countries in the best condition even in the face of a crisis for tourism, which is the third biggest sector after logistics and construction in the EU, and to form reference sets to establish how the input-output activity is provided. The findings based on the analysis of the data for three-year periods are given in Tab. 3.

Tab. 3: Efficiency Scores of European Union Countries for Three-Year Periods

Countries	2008–2010	2011–2013	2014–2016	2017–2019
Austria	1	1	1	1
Belgium	1	1	0,89466	0,667593
Bulgaria	1	1	0,884434	0,864435
Cyprus	1	1	1	1
Czechia	0,66988	0,592148	0,525005	0,521917
Denmark	1	0,991576	0,998441	1
Estonia	0,759727	0,672245	0,559139	0,518845
Finland	0,493399	0,522754	0,42031	0,44178
France	0,90213	0,890786	1	1
Germany	0,858105	0,808815	0,89654	0,852357
Greece	1	1	1	1
Hungary	0,955898	0,847108	0,781727	0,817496
Ireland	0,991038	0,883068	0,899605	0,943977
Italy	0,762939	0,737021	0,70182	0,731568
Latvia	1	1	0,99555	0,436126
Lithuania	1	0,999886	0,468286	0,460099
Luxembourg	1	1	1	1
Malta	1	1	1	1
Netherlands	0,658769	0,615547	0,657541	0,676389

Tab. 3: Continued

Countries	2008–2010	2011–2013	2014–2016	2017–2019
Norway	0,374036	0,534283	0,492297	0,489973
Poland	1	0,99697	0,976566	1
Portugal	0,987235	0,998532	1	1
Romania	0,487301	0,417487	0,385416	0,433182
Slovakia	0,701116	0,599106	0,537515	0,57774
Slovenia	0,975356	0,821321	0,662344	0,594724
Spain	1	1	1	1
Sweden	0,82938	0,894424	1	0,822572
United Kingdom	1	1	1	1

Source: Author

In Tab. 3, the findings of the output-oriented BCC DEA model are given. According to the findings, at the beginning of the 2008 global crisis, 13 countries are found to be efficient and 15 countries inefficient in tourism sector. The first period during which the number of inefficient countries increased after the 2008 economic crisis was from 2011 to 2013. As can be seen from the decline of efficient countries, the performance efficiency in tourism sector around the EU countries is clearly affected by the changing economic conditions. The second three-year period after the crisis, which is between 2014 and 2016, has the same number of efficient countries as the previous period. This situation indicates that the economic balance in the tourism sector is tried to be achieved. In the last three-year period, the number of efficient countries did not reach the level during which the crisis had just started.

Tab. 4: Ranking Result of EU Countries

Countries	2008–2010	2011–2013	2014–2016	2017–2019
Austria	1	1	1	1
Belgium	1	1	16	19
Bulgaria	1	1	17	13
Cyprus	1	1	1	1
Czechia	24	25	24	22
Denmark	1	14	11	1
Estonia	22	22	22	23
Finland	26	27	27	26
France	18	16	1	1
Germany	19	20	15	14
Greece	1	1	1	1
Hungary	17	18	18	16
Ireland	14	17	14	12
Italy	21	21	19	17
Latvia	1	1	12	27
Lithuania	1	11	26	25
Luxembourg	1	1	1	1
Malta	1	1	1	1
Netherlands	25	23	21	18
Norway	28	26	25	24
Poland	1	13	13	1
Portugal	15	12	1	1
Romania	27	28	28	28
Slovakia	23	24	23	21
Slovenia	16	19	20	20
Spain	1	1	1	1
Sweden	20	15	1	15
United Kingdom	1	1	1	1

Source: Author

As shown in Tab. 4, efficient countries whose efficiency scores are equal to 1 include Austria, Cyprus, Greece, Luxembourg, Malta, Spain and United Kingdom for all three-year periods following the 2008 crisis. In terms of the variables used, the country with the lowest efficiency score for all periods is found to be Romania. To interpret all the findings in more detail, reference sets and λ density values of EU countries according to output-oriented BCC model must be inquired.

Tab. 5: Reference Countries and λ Density Values of EU Countries for 2008–2010 Period

Countries	Refer. Country	λ	Refer. Country	λ	Refer. Country	λ	Refer. Country	λ	Refer. Country	λ
Austria	Austria	1								
Belgi	Belgium	1								
Bulgaria	Bulgaria	1								
Cyprus	Cyprus	1								
Czechia	Greece	0,315	Luxemb.	0,622	Spain	6E-02				
Denmark	Denmark	1								
Estonia	Luxemb.	0,105	Malta	0,894	Spain	2E-03				
Finland	Cyprus	0,065	Denmark	0,108	Greece	4E-02	Luxemb.	0,763	Spain	2E
France	Spain	0,784	United K.	0,216						
Germany	Spain	0,244	United K.	0,756						
Greece	Greece	1								
Hungary	Greece	0,193	Luxemb.	0,481	Malta	0,309	Spain	1,68E-02		
Ireland	Austria	0,089	Cyprus	0,722	Luxemb.	0,174	Spain	1,55E-02		
Italy	Spain	1								
Latvia	Latvia	1								
Lithuania	Lithuania	1								
Luxemb.	Luxemb.	1,								
Malta	Malta	1								
Netherlan	Luxemb.	0,671	Spain	0,145	United K.	0,185				
Norway	Austria	0,445	Belgium	0,144	Luxemb.	0,412				
Poland	Poland	1,								
Portugal	Luxemb.	0,411	Malta	0,458	Spain	0,131				
Romania	Bulgaria	0,361	Luxemb.	0,295	Malta	0,344				
Slovakia	Bulgaria	0,454	Luxemb.	0,359	Malta	0,187				
Slovenia	Luxemb.	0,287	Malta	0,700	Spain	0,012				
Spain	Spain	1,000								
Sweden	Luxemb.	0,837	Poland	5E-04	Spain	4E-02	United K.	0,122		
United K.	United K.	1,000								

Source: Author

Reference countries create reference sets which help to give suggestions, thereby allowing for the determination of the improvement percentages of the inefficient countries and λ density values are given in Tab. 5. According to the reference sets given in the table, the most referenced countries are Luxembourg and Spain, each for 12 times, at the beginning of the global economic crisis. The remaining members of the reference set are Malta (6 times), United Kingdom (4 times), Greece (3 times), Austria (2 times), Bulgaria (2 times), Cyprus (2 times). Through the countries in the reference set of an inefficient country and λ density values, input quantities to achieve to be efficient can be calculated by using actual and target values. Tab. 6 gives the projection of input values for inefficient countries to BCC model.

Tab. 6: Actual and Target Input Values of Inefficient EU Countries for 2008–2010 Period

Country	Actual Input Values				Target Input Values			
	I1	I2	I3	I4	I1	I2	I3	I4
Czechia	4941	4463666667	653273	6338592	2780	4463666667	582332	6338592
Estonia	606	682333333	48832	1459279	202	682333333	48832	1106288
Finland	2479	4417666667	217191	2344438	1116	4417666667	217191	2344438
France	25777	38738666667	5551669	37244376	21547	28558392732	3207583	37244376
Germany	38335	83556000000	3262917	25249234	26839	54481753629	3144412	25249234
Hungary	3776	2793000000	305401	3376594	1351	2793000000	305401	3376594
Ireland	2047	8437333333	211562	3743616	973	2925814440	211562	3743616
Italy	22772	28603000000	4648861	42238595	19434	18205000000	3232813	42035052
Netherlan	8350	19159000000	1203174	10302700	8350	17283321703	1087783	10302700
Norway	2505	13224333333	507597	11299400	7833667	4581333333	2505	9036031110
Portugal	4995	4004666667	468444	6719031	26179159	10230000000	2701	4004666667
Romania	9108	1760000000	302888	1362608	2931164	1452666667	1285	1760000000
Slovakia	2373	2094666667	165186	1445231	4200575	2417000000	1574	2094666667
Slovenia	981	1476666667	87527	1789643	4807530	2642000000	415	1476666667
Sweden	4539	11827666667	784202	4844906	4539	11827666667	568191	4844906

Source: Author

Target input values of the inefficient countries are calculated through density values of reference countries as given in Tab. 5. For instance, reference countries of Germany are found to be Spain with $\lambda_{26} = 0{,}244$ and United Kingdom with $\lambda_{28} = 0{,}756$ density value. Germany's target values of input variables are using the below formula to become efficient.

$$X_{Germany} = (x_{I1}, x_{I2}, x_{I3}, x_{I4})$$
$$= \{Spain's\ actual\ inputs\ values\} * 0{,}244 + \{United\ Kingdom's\ actual\ inputs\ values\} * 0{,}756$$
$$= (26839; 54481753629; 3144412; 25249234)$$

According to the calculation, to make Germany an efficient country by using inputs of tourism sector, the number of employees should be reduced from 38335 to 26839, which means that 29,99 % improvement is needed for the number of employees. Tourism expenses as the second input should also be reduced from 83556000000 to 54481753629. Thus, it can be applied − 34,80 % improvement to tourism expenses. By −3,63 % amendment of the number of beds, third input can be turned as an efficient output. Finally, it was determined that the number of tourist arrivals, which shows the tourism size of the countries, remained the same. This result coincides with the opinion that Germany's tourism size should not be reduced under the influence of the 2008 economic crisis, but should remain stable. The analyses are carried out for the output-oriented BCC model, so after the desired changes are made on the input values, a positive change will be seen in the output values. According to the calculations, with such a change, the number of nights spent by tourists will increase by 100,54 % while tourism revenues will increase by 16,54 %.

It can be said that 15 inefficient EU countries for the 2008–2010 period require some improvements to change bad input-output combination; however, the EU countries try to maintain their share of tourism sector in the world. While the effects of the economic crisis continue, the following Tab. 7 shows the extent to which countries should improve their inputs in order to strengthen their activities in the tourism sector in the second three-year period (2011–2013), unlike the beginning of the crisis.

The effects of the crisis continue to be seen in the second three-year period (2011–2013). Denmark, Lithuania and Poland, which are efficient at the onset of the crisis, are determined as ineffective countries for 2011–2013. Target input values of the inefficient countries are calculated via density values of reference countries as given in Tab. 5.

Tab. 7: Actual and Target Input Values of Inefficient EU Countries for 2011–2013 Period

Country	Actual Input Values				Target Input Values			
	I1	I2	I3	I4	I1	I2	I3	I4
Czechia	4900	4632666667	711303	7404659	2970	4632666667	541745	7404659
Denmark	2647	9837333333	421886	2235841	1797	5868464237	255311	2235841
Estonia	613	887000000	53314	1873856	224	887000000	53314	1290547
Finland	2471	5022000000	240401	2732630	1386	5022000000	240401	2732630
France	25783	42248000000	5025677	43312175	20032	24979419527	3446199	43312175
Germany	39148	86913333333	3325798	29990395	24001	43071663904	3325798	29990395
Hungary	3826	2092333333	369648	4074253	2951	2092333333	368325	4074253
Ireland	1902	6273666667	209324	2738990	1318	3435702867	209324	2738990
Italy	22452	27312000000	4744282	48820873	17731	16405000000	3414288	48517359
Lithuania	1274	951000000	58067	1117183	1274	951000000	58064	1117183
Netherlands	8307	20405666667	1274839	11905295	8307	15928858288	1221392	11905295
Norway	2574	16963666667	533210	4658077	2574	6652910772	422180	4658077
Poland	15574	8465333333	653708	4877272	3393	8465333333	507272	4877272
Portugal	4572	3943666667	488017	7722383	2458	3943666667	488017	7143589
Romania	8561	1934000000	289076	1628199	458	1934000000	116890	1628199
Slovakia	2325	2231333333	183648	1537062	683	2231333333	123176	1537062
Slovenia	922	1443666667	101336	2090844	464	1443666667	101336	1883880
Sweden	4663	14428666667	796462	4993098	4663	11410185856	631260	4993098

Source: Author

Tab. 8: Actual and Target Input Values of Inefficient EU Countries for 2014–2016 Period

Country	Actual Input Values				Target Input Values			
	I1	I2	I3	I4	I1	I2	I3	I4
Belgium	4561	17304333333	367709	7907867	1805	3953511871	367709	5934155
Bulgaria	3010	1226333333	321662	3014294	597	1226333333	181188	3014294
Czechia	5052	4941666667	713986	8708079	2921	4941666667	545010	8708079
Denmark	2711	9511666667	418101	2621345	1442	5270741137	223875	2621345
Estonia	637	1123000000	58546	1989668	296	1123000000	58546	1495390
Finland	2444	5097333333	251973	2714066	1405	4893661114	208979	2714066
Germany	40450	83558333333	3331624	34365159	17390	28527024305	3331624	34365159
Hungary	4221	2010666667	440823	4949368	1081	2010666667	359787	4949368
Ireland	2059	6116666667	201791	2694034	1155	3715764722	180914	2694034
Italy	22501	26092666667	4890426	54479663	20011	26092666667	3578607	54369866
Latvia	391	677333333	42656	1493145	886	675132458	42656	1492662
Lithuania	1338	1007333333	74571	1421071	235	1007333333	56148	1421071
Netherland	8327	19678666667	1371814	14920501	7459	16328052043	1341635	14920501
Norway	2635	16302000000	555086	5359667	2635	7542418279	476036	5359667
Poland	16047	8028000000	717829	5846233	2884	8028000000	519761	5846233
Romania	8533	2436666667	320312	2205820	621	2436666667	124579	2205820
Slovakia	2426	2264666667	185647	1718844	429	2264666667	93483	1718844
Slovenia	916	1341666667	109770	2677996	748	1341666667	109770	2517584

Source: Author

Tab. 9: Actual and Target Input Values of Inefficient EU Countries for 2017–2019 Period

Country	Actual Input Values				Target Input Values			
	I1	I2	I3	I4	I1	I2	I3	I4
Belgium	4742	17581333333	386404	8948917	2790	4217011082	386404	8948917
Bulgaria	3179	1796000000	341942	3877780	645	1796000000	183311	3877780
Czechia	5273	5775333333	733504	10554121	3097	5775333333	562018	10554121
Estonia	665	1423000000	61271	2183370	326	1423000000	61271	1772552
Finland	2526	5786666667	258617	3231552	1595	5786666667	235757	3231552
Germany	41993	92222000000	3482020	38498879	18608	33581061393	3482020	38498879
Hungary	4468	2616333333	417448	5921259	1726	2616333333	242422	5921259
Ireland	2258	7437666667	205235	3246578	1356	4654628030	205235	3246578
Italy	23199	29404000000	5108933	62909538	20251	29404000000	3579053	62909538
Latvia	905	750333333	54271	1883430	295	750333333	54271	1871267
Lithuania	1369	1301000000	93022	1755502	312	1301000000	58804	1755502
Netherlands	8795	22310000000	1396145	18944639	7849	16021238110	1396145	18944639
Norway	2682	16802666667	581610	5920859	2682	8034193207	455756	5920859
Romania	8680	5233666667	346181	2736204	1282	5233666667	199651	2736204
Slovakia	2560	2540000000	194216	2250868	435	2540000000	115772	2250868
Slovenia	974	1643000000	151864	3926943	950	1643000000	151864	3926943
Sweden	5084	15709333333	816188	7155144	4833	13297832556	816188	7155144

Source: Author

The number of inefficient countries increased with the addition of Belgium and Bulgaria in the period between 2014 and 2016. According the numbers given in Tab. 8, these two countries used their tourism input efficiently in the previous periods, increased their tourism expenditures and hired a greater number of employees to prevent from being affected by the crisis. Some problems occurred in the transformation of this increased input into output and caused the inputs to not to be used efficiently. France failed to handle it efficiently in the first two periods of the crisis but later achieved to be an efficient country in the subsequent periods by considering the input-output activity of the reference countries such as Spain and United Kingdom.

Tab. 9 presents the list of countries which failed to use the tourism sector efficiently due to different factors during the period when the impact of the crisis lessened. Italy can be seen as the most surprising one among these countries. However, this lack of success can be linked to the fact that Italy's economy is already in a vulnerable state despite the number of incoming tourists and high revenues.

Conclusion

In this research, tourism sector efficiency for EU countries that suffer from the 2008 global economic crisis is analyzed sequentially by variables as the number of employees, number of beds, tourism expenses, number of tourist arrivals, number of nights stayed, and tourism revenue. The analysis method is an output-oriented BCC DEA model, which helps to make satisfactory observations about the inputs how to make an improvement to reach highly effective outputs results. Moreover, this model gives the results of variable returns to scale model and is used with the data set of three-year periods (2008–2010, 2011–2013, 2014–2016, 2017–2019). Considering that the worldwide international tourist arrival was 1.4 million in 2019, the international tourism receipts was 1.466 billion USD dollars before the negative impact of COVID-2019, the tourism market share and efficiency of the tourism potential of European Union countries is very important. Therefore, it is essential to discuss the tourism efficiency of EU countries after the 2008 economic crisis. According to the results of the BCC DEA model at the first three-year period (2008–2010), 13 countries (United Kingdom, Austria, Belgium, Bulgaria, Cyprus, Spain, Denmark, Poland, Malta, Luxembourg, Lithuania, Greece, and Latvia) are determined to be efficient countries as DMUs. During the second three-year period (2011–2013), the number of efficient countries fell to 10 countries (United Kingdom, Austria, Belgium, Bulgaria, Cyprus, Spain, Malta, Luxembourg, Latvia, Greece). This decrease in the number of efficient countries shows that the impact of the economic crisis on the tourism sector has continued. After examining the third period (2014–2016), it is observed that the number of efficient countries was the same as the second period. Accordingly, the efficient countries are United Kingdom, Austria, Sweden, Spain, Cyprus, Portugal, Malta, Luxembourg, Greece, and France. The countries identified as relatively efficient during the last period (2017–2019) were United Kingdom, Austria, Spain, Portugal, Cyprus, Poland, Denmark, Malta, Luxembourg, France and Greece.

The striking aspect of this analysis is the presence of countries which ran the input of the tourism sector efficiently throughout the 2008 economic crisis and had the input-output balance to set an example for those inefficient countries. One of these leading countries is the United Kingdom, which officially exited the EU in early 2020. It is followed by Austria, Spain, Malta, Luxembourg and Greece. These countries considered the negative impact of the crisis on the tourism sector and continued to efficiently transform their input into output (European Commission, 2011).

France is one of the main countries which are most affected by the crisis in terms of tourism revenues. Hence, this explains why it was not included in the category of efficient countries in the two three-year periods of the crisis. In the subsequent periods, France overcame the impact of the crisis and has in recent years become one of the efficient countries that can take the input recoveries of Spain and the United Kingdom as reference examples. It is essential to use different variables for a detailed evaluation of the efficient use of inputs. According to the reference sets' results for three-year periods, many countries faced difficulties in the tourism market. Some of these EU countries that include Czechia, Estonia, Finland, Germany, Hungary, Ireland, Italy, Netherland, Norway, Romania, Slovakia, Slovenia seemed to have used their input variables inefficiently for the periods after the economic crisis. At this point, it can be stated that these countries are required to decrease redundant tourism expenditures, the numbers of employees and try to follow the examples of efficient tourism management throughout the economic crisis.

It can be suggested that if long-term success is desired, EU countries that have not received the desired share in the tourism sector should make a detailed examination of the tourism sector. The most important factors that make EU countries more preferable are not only the number of beds, the number of employees and the number of nights that tourists spend, but also the touristic historical places, nature, and hospitality of people living in that travel destination. In addition to all these irreplaceable things, arrangements can be made such as educating the employees instead of dismissing them and providing the comfort of existing beds instead of increasing the bed capacity. These regulations can help gain a good place in the tourism sector competition.

Bibliography

Atan, M. and Arslantürk, Y. (2015), Dünya Ülkelerinin Turizm Potansiyelinin Etkinliği, *Gazi Journal of Economics and Business*, 1 (1), pp. 59–76.

Bakırcı, F. (2006), *Üretimde Etkinlik Ölçümü ve Veri Zarflama Analizi Teori ve Uygulama*, Tokat: Atlas Yayınları.

Banker, Rajiv D., Charnes, A., and Cooper, W. W. (1984), Some Models for Estimating Technical and Scale Inefficiencies in Data Envelopment Analysis, *Management Science*, 30 (9), pp. 1078–2092.

Barišić, P. and Cvetkoska, V. (2020), Analyzing the Efficiency of Travel and Tourism in the European Union. In N. Mladenović, A. Sifaleras and M. Kuzmanović (Eds.), *Advances in Operational Research in the Balkans*, Springer Proceedings in Business and Economics, Springer, pp. 167–186.

Bayrak, R. and Bahar, O. (2018), Economic Efficiency Analysis of Tourism Sector in OECD Countries: An Emprical Study with DEA, *Uluslararası İktisadi ve İdari İncelemeler Dergisi*, 20, pp. 83–100.

Berberoğlu, B. (2011), 2008 Global Krizinin Türkiye ve Avrupa Birliği'ndeki Etkilerinin Kümeleme Analizi İle İncelenmesi, *Anadolu Üniversitesi Sosyal Bilimler Dergisi*, 11 (1), pp. 105–130.

Chaabouni, S. (2019), China's Regional Tourism Efficiency: A Two-Stage Double Bootstrap Data Envelopment Analysis, *Journal of Destination Marketing & Management*, 11, pp. 183–191.

Charnes, A., Cooper, W. W., and Rhodes, E. L. (1978), Measuring the Efficiency of Decision Making Units, *European Journal of Operational Research*, 2 (6), pp. 429–444.

Cooper, W. W., Seiford, L. M., and Kaoru, T. (2007), *Data Envelopment Analysis: A Comprehensive Text with Models, Applications, References and DEA- Solver Software*, USA: Kluwer Academic Publishers. Second edition.

Cracolici, M., Francesca, N. P., and Rietveld, P. (2008), Assessment of Tourism Competitiveness by Analysing Destination Efficiency, *Tourism Economics*, 14 (2), pp. 325–342.

Demir, E. Ö. (2010), *Turizm Sektöründe Kriz ve Kriz Yönetimi* (Expertise Thesis), Avaiable from Kültür ve Turizm Bakanlığı Tanıtma Genel Müdürlüğü, Ankara.

European Commission (2011), *European Economic Forecast*, Luxembourg: Publications office of the European Union.

Furceri, D. and Mourougane, A. (2012), The Effect of Financial Crises on Potential Output: New Empirical Evidence from OECD Countries, *Journal of Macroeconomics*, 34, pp. 822–832.

Hadad, S., Hada, Y., Malul, M., and Rosenboi, M. (2012), The Economic Efficiency of the Tourism Industry: A Global Comparison, *Tourism Economics*, 18, pp. 931–940.

Kocakoç, İ. D. (2003), Veri Zarflama Analizi'ndeki Ağırlık Kısıtlamalarının Belirlenmesinde Analitik Hiyerarşi Sürecinin Kullanımı, *Dokuz Eylül Üniversitesi İİBF Dergisi*, 18 (2), pp. 1–12.

Liu, J., Ding, F.-Y., and Lall, V. (2000). Using Data Envelopment Analysis to Compare Suppliers for Supplier Selection and Performance Improvement. *Supply Chain Management: An International Journal*, 5 (3), 143–150.

Man, D. and Zhang, H. (2015), *The Study of DEA Application in Tourism City: A Case for Members of the World Tourism City Federation in China*. In Z. Zhang, Z. Max Shen, J. Zhang, and R. Zhang (Eds.), LISS 2014, Springer, Berlin, Heidelberg, pp. 831–835.

Selamoğlu, A. (2015), 2008 Krizi ve Avrupa Birliği Yeni Yönetişim Kıskacında Endüstri İlişkileri, *Sosyal Siyaset Konferansları*, 68 (1), pp. 25–61.

Shurak, T. (2001), Tourism Policy, *Polityka Gospodarcza*, 5 (6), pp. 274–283.

Soysal-Kurt, H. (2017), Measuring Tourism Efficiency of European Countries by Using Data Envelopment Analysis, *European Scientific Journal*, 13, pp. 31–49.

Tone, K. (1996), A Simple Characterization of Returns to Scale in DEA, *Journal of the Operations Research Society of Japan*, 39 (4), pp. 604–613.

United Nations World Trade Organization-UNWTO (2016), UNWTO Tourism Higlights, http://www.e-unwto.org/doi/pdf/10.18111/9789284418145 (Access Date: December 7, 2021).

Yılmaz, B. E. (2013), Reflections of The Global Economic Crisis on the Countries of PIIGS and Turkey's Macroeconomic Variables, *Marmara Üniversitesi İ.İ.B.F. Dergisi*, 34 (1), pp. 229–252.

Yi, T. and Liang, M. (2014), Evolutional Model of Tourism Efficiency Based on the DEA Method: A Case Study of Cities in Guangdong Province. China, *Asia Pacific Journal of Tourism Research*, 20, pp. 789–806.

Nesrin ÖZKAN

The Impact of COVID-19 on City Indices in Turkey: An Event Study Analysis

The study investigates the impact of COVID-19 on city indices in Turkey. The motivation underlying this research is to reveal whether all city indices are affected from pandemic equally or there is a difference among them. For that purpose, 12 city indices' returns are analyzed by event study. Event dates are chosen on 11th of March, 2020 when World Health Organization (WHO) announced the pandemic and first case detected in Turkey, and 10th of April, 2020 when the declaration of curfew imposed for 31 cities in Turkey. According to the findings, the pandemic announcement causes negative abnormal returns in city indices, especially in Antalya and Denizli. Antalya city indice records the highest losses in event windows. The pandemic disease has great impact on the economy of the city because the city is one of the top tourism destinations in the world to visit. The returns vary as per the event windows, but the highest loss is observed in (−10; 10) event window. On the other hand, the first curfew has no impact on city indices, contrarily.

1 Introduction

COVID-19 burst out in the Wuhan City of China nearly at the end of 2019, and have spreaded the world rapidly. No longer than emerged, it has been declared a pandemic on 11 March of 2020 by World Health Organization. Although the curfew and quarantine precautions, millions of people have been affected from the pandemic. As of 27 August 2020, 24.021.218 confirmed cases of COVID-19 have been reported to World Health Organization, and the number consists of 821.462 deaths (WHO, 2020).

The pandemic did not only affect the health of communities but also the global economy, financial markets and businesses. Since unexpected events breed the uncertainty in financial markets, huge losses were recorded in indexes of international markets such as Dow Jones, S&P, Nasdaq, Dax, Nikkei 225. BIST 100 index recorded loss as well and dramatically dropped to 820 points due to the COVID-19. Besides that, the COVID-19 affects the businesses on sectorial basis. In order to measure the effect of COVID-19 outbreak on sectors, Göker et al. (2020) investigated 26 sector indices in Borsa İstanbul. The authors searched how the sectors were affected from outbreak, and whether there was a difference among sectors. The daily closing values of 26 indices were used in analysis, and the data covered the dates from 2 January, 2019 to 9 April, 2020. The event study approach was employed, and 8 event windows were determined as (−20; +20), (−10; +10), (−5; +5), (−2; +2), (−1; +1), (0; 0), (0; 10), and (0; 20). The date of 11 March, 2020 was chosen as the event date. According to the results, most of 26 indices recorded loss, but the highest losses were realized in Sports, Tourism and Textile indices. On the other hand, Food Beverage, Chemical Petrol Plastics, Mining, Industrials, and Wood Paper Printing indices had positive CAARs in analysis period. Kılıç (2020) conducted the similar analysis. Four event dates were determined as 11 January, 2020 when the first death was recorded due to COVID-19 virus, 11 February, 2020 when WHO declared the COVID-19 is a new type of virus; 11 March, 2020 when WHO announced the pandemic and first case detected in Turkey; and 11 April 2020 when the declaration of curfew was imposed for 31 cities in Turkey. The findings showed that Tourism and Textile sectors had the highest negative abnormal returns. On the other hand, Food Beverage sector exhibited a positive and statistically significant return. It was further documented that the substantial losses were on 11 March, 2020 specifically and 11 April, 2020 did not cause abnormal returns.

The recent studies on economic and financial consequences of pandemic vary from the impact of pandemic on economic performance and macroeconomic indicators to financial indexes in literature. Yan (2020) analyzed the abnormal returns in Chinese stock market in the course of COVID-19 outbreak by event study. The author reported that COVID-19 caused big moves on prices of stocks. Sansa (2020) investigated the relationship between confirmed COVID-19 cases, and the Dow Jones, Shanghai Stock Exchange markets between 11 March, 2020 and 25 March, 2020. The results disclosed the positive relation. Şenol and Zeren (2020) examined the relationship between COVID-19 cases and death numbers with Morgan Stanley Capital International and G7 indices between 21 January, 2020 and 7 April, 2020. Fourier cointegration test revealed a long-term relationship between global markets and COVID-19. Zeren and Hızarcı (2020) investigated the cointegration between stock market indexes of China, South Korea, France, Italy, Germany and Spain and the daily cases and deaths, and reported COVID-19 cases had a cointegration with the China, South Korea and Spain indices. Açıkgöz and Günay (2020) discussed the possible economic and political outcomes of COVID-19 on post-pandemic world. Alber (2020) analyzed the relationship between COVID-19 cases, deaths, new cases and past COVID-19 deaths with the six countries' stock market indexes where the COVID-19 had the worst situation, and found that these indexes had been affected more from cases than deaths. Sarı and Kartal (2020) found the number of COVID- 19 cases affected gold prices and VIX index. Chowdhury and Abedin (2020) examined the relationship between COVID-19 and the US. stock market via GARCH, VAR and ESM methods, and reported that confirmed and death cases were negatively related with the stock market in the US. Aawadhi et al. (2020) reported COVID-19 daily growth in total confirmed cases and total cases of death had a negative impact on the Chinese stock market. Albulescu (2020a) investigated the impact of new confirmed cases and death ratio on the financial markets' volatility index, and found a positive relation between death ratio and volatility index (VIX). Albulescu (2020b) further searched COVID-19 daily reported new cases and crude oil prices, and reported that the new cases had a marginal negative impact on the crude oil prices. Mazur et al. (2020) examined the effect of COVID-19 outbreak on Standard and Poor's (S&P) 1.500 firms. The stocks of natural gas, food, healthcare, and software sectors exhibited positive returns, but the stocks of petroleum, real estate, entertainment, and hospitality sectors were affected negatively. Özkan and Ünlü (2021) investigated the cointegration between regional COVID-19 cases, gold prices, Euro exchange rate, and BIST city indices of İstanbul, Ankara and İzmir. The research adopted ARDL Bound Test in order to test the short-run and long-run relations. The authors documented a long-run relationship among BIST İstanbul, BIST İzmir indices and COVID-19 cases, gold prices and Euro exchange rate. Besides that, Ali et al. (2020), Feng et al. (2020), Akhtaruzzaman et al. (2020), Ruiz et al. (2020), Baker et al. (2020), Luo and Tsang (2020), Zhang et al. (2020), Beck (2020) and Makridlis and Hartley (2020) are the other researchers investigating the effects of COVID-19 pandemic on global economy and financial markets.

As to Bayramoğlu and Pekkaya (2010: 201–210), city indices are the indicator of regional financial performances, and they are helpful to decide whether to invest the city. Besides, these indices reveal the regional potentials of city, and enable to the analysis in micro basis. This analysis may provide regional information to the investors and opportunity to the investors to make comparisons of cities' economic performances. For that purpose, it has been calculating indices for 153 provinces in the US., and they became the model for Turkey to start calculating the city indices in Borsa İstanbul. Borsa İstanbul has been calculating the city indices since the beginning of 2009 and the closing value of BIST 100 Index on the date 31 December 2008 is taken as a basis. The indices are calculating for 12 cities those are Adana (XSADA), Ankara (XSANK), Antalya (XSANT), Balıkesir (XSBAL), Bursa (XSBUR), Denizli (XSDNZ), İstanbul (XSIST), İzmir (XSIZM), Kayseri (XSKAY), Kocaeli (XSKOC), Konya (XSKON) and Tekirdağ (XSTKR). To have a city indice, the number of companies traded in BIST must reach to five for the city. The equity of companies ought to have traded on Stars, Main and Emerging Companies markets. In calculation and inclusion of city indices, there are some basic rules for companies as follows (BIST, 2020):

1. The financials except holdings and retail companies are not included to the indice.
2. For the production companies, minimum 50 % of the production ought to take place in the city.
3. For communication companies, construction companies and holdings, the location of registered office is taken into account.
4. For service companies (except communication and construction companies and holdings), minimum 50 % of the operating income ought to have derived in the city. If there is no city where minimum 50 % of the production or operating income is realized, then the city where the registered office is located is taken into account.

The purpose of calculation the city indices is to reflect the shares' performance of the companies in the city which enables the creation of a variety of new financial products such as Exchange Traded Funds (ETFs) and so investors could be able to invest in a city through ETFs.

City indices enable to permanently analyze the latest news and developments about the issues or problems of cities (Bayramoğlu & Pekkaya, 2010: 200). Since they are created to measure price and return performances of shares of companies, it gives possibility to measure the economic performance of the cities, and compare it with one another (Bayramoğlu & Pekkaya, 2010: 210). Thus, the city indices might inherit the information about the impact of COVID-19 on city's economic performance. Relevant studies focus more on the impact of COVID-19 on sector indices, this study linkup the performance of city indices with COVID-19 announcements. The rest of the paper is organized as follows. Section 2 describes the data about the city indices, and research questions of the study. Section 4 explains the methodology thoroughly. Section 5 consists of the empirical results. Section 6 covers the conclusion, and summary of main findings.

2 Data

In this study, the aim is to unveil the impact of COVID-19 on BIST City Indices in Turkey. To that aim, 12 city indices' closing prices are employed between 11 February, 2019 and 12 May, 2020. The data are collected from Borsa İstanbul DataStore, and BIST 100 Index is used as a market proxy. Tab. 1 displays the properties about 12 city indices.

Tab. 1: The List of Companies in City Indices

City Indice	Indice Code	Number of Companies	Indice Starting Date	Initial Value
BIST ADANA	XSADA	3	31.12.2008	28.864,07
BIST ANKARA	XSANK	21	31.12.2008	28.864,07
BIST ANTALYA	XSANT	5	31.12.2008	28.864,07
BIST BALIKESİR	XSBAL	6	12.5.2011	66.535.13
BIST BURSA	XSBUR	17	31.12.2008	28.864,07
BIST DENİZLİ	XSDNZ	4	6.7.2012	61.972,14
BIST İSTANBUL	XSIST	100	31.12.2008	28.864,07
BIST İZMİR	XSIZM	29	31.12.2008	28.864,07
BIST KAYSERİ	XSKAY	9	31.12.2008	28.864,07
BIST KOCAELİ	XSKOC	19	31.12.2008	28.864,07
BIST KONYA	XSKON	5	4.12.2012	75.522,65
BIST TEKİRDAĞ	XSTKR	4	31.12.2008	28.864,07

This study emerges by searching the answer of a question. The question is which city indice is affected from the announcements related to COVID-19, and if there is a difference among them. Or more precisely, how the city indices reacted at dates of pandemic declaration of WHO and first case detected in Turkey, and curfew declaration in Turkey. The curfew was imposed in 31 cities[1] totally, and the indices have already been calculating 12 out of 31. In other words, the curfew enforced cities covers the city indices calculated in Borsa İstanbul. The event dates are determined as 11 March of 2020 and 10 April of 2020. The data are processed in Excel and analyzed by using Stata 15.0.

3 Methodology

In this research, the event study approach is employed which is commonly used for measuring the effects of economic, financial, political even social events (e.g., terrorist attacks, mergers and acquisitions, dividend payouts, initial public offerings, crisis, pandemics) on the value of the stock. The logic behind the event study is linked with the Efficient Market Hypothesis of Fama

---

1 Those are Adana, Ankara, Antalya, Aydın, Balıkesir, Bursa, Denizli, Diyarbakır, Erzurum, Eskişehir, Gaziantep, Hatay, İstanbul, İzmir, Kahramanmaraş, Kayseri, Kocaeli, Konya, Malatya, Manisa, Mardin, Mersin, Muğla, Ordu, Sakarya, Samsun, Şanlıurfa, Tekirdağ, Trabzon, Van and Zonguldak.

(1970). According to Fama (1970: 383–416), the markets are categorized into three in terms of processing the information. They are weak-form market efficiency where all historical prices reflect the stock prices; semi-strong market efficiency where it additionally reflects the public information available, and strong-form market efficiency where the stock prices reflect all information whether it is publicly known or not. In order to test the efficiency of markets, efficiency tests are used. The weak-form efficiency could be measured by return predictability tests. The event studies are used to measure the semi-strong market efficiency, and private information tests are used for measuring the strong-form market efficiency (Fama, 1991: 1576–1577). In light of the statement, this study attempts to measure the semi-strong form of market efficiency by event study.

The event windows are selected as (−1; 1), (−2; 2), (−5; 5), (−10; 10), and (−20; 20). A typical event study analysis consists of three main parts which are pre-event interval, event date, and post-event interval. It is shown detailed in timeline in Fig. 1 as follows:

Fig. 1: Timeline for Event Study
Source: The author.

In timeline, the point 0 shows the event day, and the interval between T-2 and T-1 is called the estimation window. The interval of T-1 and T1 is the event window, and the interval between T1 and T2 is the post event window. In event studies, the selection of model is crucial. The market and risk-adjusted models have found better than the others since they are lacking in market or risk adjustment. Furthermore, the market model is frequently used one in event studies. Other complex models' performance is fairly the same as market model (Armitage, 1995). Therefore, the market model is preferred by using the Ordinary Least Squares (OLS) method in our analysis. The steps of the market model could be expressed by the equations from 1 to 7 as written below. First of all, the market return is calculated from daily closing values of BIST 100 index.

$$R_{m,t} = \frac{P_{m,t} - P_{m,t-1}}{P_{m,t-1}} \qquad (1)$$

where,
$R_{m,t}$ = The return of market in time t
$P_{m,t}$ = The closing value of market index in time t
$P_{m,t-1}$ = The closing price of market index in time t-1
The daily indice returns are calculated in the same way as follows:

$$R_{i,t} = \frac{C_{i,t} - C_{i,t-1}}{C_{i,t-1}} \qquad (2)$$

where,
$R_{i,t}$ = The return of indice i in time t
$C_{i,t}$ = The closing value of indice i in time t
$C_{i,t-1}$ = The closing price of indice i in time t-1
After the calculation of market and indice returns, the market model is constructed as in Equation 3.

$$R_{i,t} = \alpha_i + \beta_i R_{mt} + \varepsilon_{it} \qquad (3)$$

where,
α_i = Intercept term
$R_{i,t}$ = The daily return of the indice i in time t
β_i = Coefficient or the slope of regression line
R_{mt} = The daily return of market in time t
ε_{it} = Error term
The expected returns are expressed in Equation 4 below

$$E(R_{i,t}) = \alpha + \beta_i (R_{m,t}) \qquad (4)$$

The abnormal returns are calculated as follows:

$$AR_{i,t} = R_{i,t} - \alpha - \left[\beta_i(R_{m,t})\right] \quad (5)$$

where,
$AR_{i,t}$ =Abnormal return of indice i on time t
For each event window, AAR_t is calculated as seen as below

$$AAR_t = \frac{1}{n}\sum_{i=1}^{n} AR_{it} \quad (6)$$

where,
AAR_t = Average abnormal return for all indices in time t
n = Number of indices
Finally, the cumulative average abnormal returns are calculated as follows:

$$CAAR(t_1,t_2) = \frac{1}{n}\sum_{i=1}^{n} CAAR(t_1,t_2) \quad (7)$$

The statistical tests are conducted to control whether abnormal returns (ARs) are statistically different from zero. The normality assumption claims that ARs are normally distributed and heteroscedastic across the series. The tests used are Patell's (1976) standardized residual test and Boehmer et al. (1991) standardized cross-sectional test are parametric tests. AdjPatell is the test of Patell's (1976) with the Kolari and Pynnönen (2010) adjustment. KP test is the BMP test that is corrected for the cross-sectional correlation of ARs. The test proposed by Wilcoxon (1945) is a nonparametric signed rank test and generalized rank test of GRANK is run to assess the statistical significance of ARs.

4 Findings

The results for 12 indices are tabulated for the event windows of (−1; 1), (−2; 2), (−5; 5), (−10; 10) and (−20; 20). Tab. 2 reported cumulative average abnormal returns (CAARs) for the event window (−1; 1).

Tab. 2: Cumulative Average Abnormal Returns for Event Window (−1; 1)

CAAR [−1,1]							
City Indice	CAAR	Patell Z	Adj. Patell Z	BMP	KP	Wilcoxon	Grank
BIST ADANA	−11.38***	−11.38***	−11.38***	−11.38***	−11.38***	−11.38***	−11.38***
BIST ANKARA	−4.03***	−4.03***	−4.03***	−4.03***	−4.03***	−4.03***	−4.03***
BIST ANTALYA	−20.76***	−20.76***	−20.76***	−20.76***	−20.76***	−20.76***	−20.76***
BIST BALIKESİR	−12.14***	−12.14***	−12.14***	−12.14***	−12.14***	−12.14***	−12.14***
BIST BURSA	−3.80*	−3.80*	−3.80*	−3.80*	−3.80*	−3.80*	−3.80*
BIST DENİZLİ	−21.28***	−21.28***	−21.28***	−21.28***	−21.28***	−21.28***	−21.28***
BIST İSTANBUL	−4.27***	−4.27***	−4.27***	−4.27***	−4.27***	−4.27***	−4.27***
BIST İZMİR	−8.77***	−8.77***	−8.77***	−8.77***	−8.77***	−8.77***	−8.77***
BIST KAYSERİ	−11.98***	−11.98***	−11.98***	−11.98***	−11.98***	−11.98***	−11.98***
BIST KOCAELİ	−4.72**	−4.72**	−4.72**	−4.72**	−4.72**	−4.72**	−4.72**
BIST KONYA	−19.41***	−19.41***	−19.41***	−19.41***	−19.41***	−19.41***	−19.41***
BIST TEKİRDAĞ	8.11**	8.11**	8.11**	8.11**	8.11**	8.11**	8.11**
Pos: Neg	01:11						

Notes: All values are expressed in percentage (%). The symbols of ***, **, and * denote significance levels at %1, %5, and %10.

In general, negative abnormal returns are observed in city indices, except BIST Tekirdağ. The CAARs realize at the lowest in BIST Antalya and BIST Balıkesir indices, respectively. In event window of (−1; 1), the loss is approximately 21 % for BIST Antalya. The main reason could be attributed to the impact of COVID-19 on Tourism Industry. The result gives the signal how Tourism sector was affected from the pandemic. When we take a closer look the companies listed in BIST Denizli indice, we could not see any specific sector dominated to the indice return because all companies listed in indice are almost from different sectors. So that it seems that the tourism sector dominates the returns but it is specific to BIST Antalya city indice, only.

On the other hand, the abnormal returns for BIST Bursa are significant at 0.10 level whereas they are significant for BIST Kocaeli and BIST Tekirdağ at 0.05 level. The statistical tests show that stock returns' reaction to 11 March, 2020 is statistically significant almost all at the level of 0.01 according to all parametric and non-parametric tests. Tab. 3 provides the CAARs for event window (−2; 2).

In Tab. 3, the results are pretty much the same as the event window of (−1; 1). It seems as though the loss is getting larger when the event window extends. BIST Antalya and BIST Denizli own the highest negative abnormal returns, and the values are statistically significant at 0.01 level. BIST Tekirdağ is the only indice providing the positive return but is not significant. The loss of BIST Antalya is nearly 25 % in the event window and BIST Denizli records second the highest loss with 24 %. BIST Tekirdağ again presents negative abnormal return but insignificant. All parametric and non-parametric tests support the significant effect of 11 March, 2020 on returns.

Tab. 4 reported the CAARs for the event window (−5; 5) The highest loss in CAAR is observed in BIST Antalya with 38.95 % loss, and BIST Denizli follows with 37.09 % loss. As it could be easily observed, the loss is getting larger by the event window extends. Contrarily to previous event windows of (−1; 1) and (−2; 2), there is no positive return observed in indices. The lowest negative abnormal return is in BIST İstanbul with 6.71 %. The CAARs over the event window with statistical tests indicate the significant effect of 11 March, 2020 on returns.

In Tab. 5, the highest loss realizes again in BIST Antalya indice with 52.15 %. It is recorded the highest loss across the indices, and event windows. BIST Tekirdağ and BIST Bursa indices present positive return, but both are statistically insignificant. On the other hand, the negative returns are higher in comparison with the other event windows. Thus, the CAARs record the highest losses in this event window of (−10; 10).

Tab. 3: Cumulative Average Abnormal Returns for Event Window (−2; 2)

CAAR [−2,2]							
City Indice	CAAR	Patell Z	Adj. Patell Z	BMP	KP	Wilcoxon	Grank
BIST ADANA	−11.52***	−11.52***	−11.52***	−11.52***	−11.52***	−11.52***	−11.52***
BIST ANKARA	−4.56**	−4.56**	−4.56**	−4.56**	−4.56**	−4.56**	−4.56**
BIST ANTALYA	−24.89***	−24.89***	−24.89***	−24.89***	−24.89***	−24.89***	−24.89***
BIST BALIKESİR	−15.31***	−15.31***	−15.31***	−15.31***	−15.31***	−15.31***	−15.31***
BIST BURSA	−9.78***	−9.78***	−9.78***	−9.78***	−9.78***	−9.78***	−9.78***
BIST DENİZLİ	−23.87***	−23.87***	−23.87***	−23.87***	−23.87***	−23.87***	−23.87***
BIST İSTANBUL	−3.70***	−3.70***	−3.70***	−3.70***	−3.70***	−3.70***	−3.70***
BIST İZMİR	−10.02***	−10.02***	−10.02***	−10.02***	−10.02***	−10.02***	−10.02***
BIST KAYSERİ	−14.63***	−14.63***	−14.63***	−14.63***	−14.63***	−14.63***	−14.63***
BIST KOCAELİ	−5.62**	−5.62**	−5.62**	−5.62**	−5.62**	−5.62**	−5.62**
BIST KONYA	−20.84***	−20.84***	−20.84***	−20.84***	−20.84***	−20.84***	−20.84***
BIST TEKİRDAĞ	2.60	2.60	2.60	2.60	2.60	2.60	2.60
Pos: Neg	01:11						

Notes: All values are expressed in percentage (%). The symbols of ***, **, and * denote significance levels at %1, %5, and %10.

Tab. 4: Cumulative Average Abnormal Returns for Event Window (−5; 5)

CAAR [−5,5]							
City Indice	CAAR	Patell Z	Adj. Patell Z	BMP	KP	Wilcoxon	Grank
BIST ADANA	−23.73***	−23.73***	−23.73***	−23.73***	−23.73***	−23.73***	−23.73***
BIST ANKARA	−8.04***	−8.04***	−8.04***	−8.04***	−8.04***	−8.04***	−8.04***
BIST ANTALYA	−38.95***	−38.95***	−38.95***	−38.95***	−38.95***	−38.95***	−38.95***
BIST BALIKESİR	−20.81***	−20.81***	−20.81***	−20.81***	−20.81***	−20.81***	−20.81***
BIST BURSA	−23.16***	−23.16***	−23.16***	−23.16***	−23.16***	−23.16***	−23.16***
BIST DENİZLİ	−37.09***	−37.09***	−37.09***	−37.09***	−37.09***	−37.09***	−37.09***
BIST İSTANBUL	−6.71***	−6.71***	−6.71***	−6.71***	−6.71***	−6.71***	−6.71***
BIST İZMİR	−20.30***	−20.30***	−20.30***	−20.30***	−20.30***	−20.30***	−20.30***
BIST KAYSERİ	−22.38***	−22.38***	−22.38***	−22.38***	−22.38***	−22.38***	−22.38***
BIST KOCAELİ	−21.44***	−21.44***	−21.44***	−21.44***	−21.44***	−21.44***	−21.44***
BIST KONYA	−35.82***	−35.82***	−35.82***	−35.82***	−35.82***	−35.82***	−35.82***
BIST TEKİRDAĞ	−20.91***	−20.91***	−20.91***	−20.91***	−20.91***	−20.91***	−20.91***
Pos: Neg	00:12						

Notes: All values are expressed in percentage (%). The symbols of ***, **, and * denote significance levels at %1, %5, and %10.

Tab. 5: Cumulative Average Abnormal Returns for Event Window (−10; 10)

CAAR [−10,10]							
City Indice	CAAR	Patell Z	Adj. Patell Z	BMP	KP	Wilcoxon	Grank
BIST ADANA	−23.16***	−23.16***	−23.16***	−23.16***	−23.16***	−23.16***	−23.16***
BIST ANKARA	1.06	1.06	1.06	1.06	1.06	1.06	1.06
BIST ANTALYA	−52.15***	−52.15***	−52.15***	−52.15***	−52.15***	−52.15***	−52.15***
BIST BALIKESİR	−26.49***	−26.49***	−26.49***	−26.49***	−26.49***	−26.49***	−26.49***
BIST BURSA	−22.40***	−22.40***	−22.40***	−22.40***	−22.40***	−22.40***	−22.40***
BIST DENİZLİ	−36.88***	−36.88***	−36.88***	−36.88***	−36.88***	−36.88***	−36.88***
BIST İSTANBUL	−4.82**	−4.82**	−4.82**	−4.82**	−4.82**	−4.82**	−4.82**
BIST İZMİR	−12.54***	−12.54***	−12.54***	−12.54***	−12.54***	−12.54***	−12.54***
BIST KAYSERİ	−24.63***	−24.63***	−24.63***	−24.63***	−24.63***	−24.63***	−24.63***
BIST KOCAELİ	−15.44***	−15.44***	−15.44***	−15.44***	−15.44***	−15.44***	−15.44***
BIST KONYA	−30.06***	−30.06***	−30.06***	−30.06***	−30.06***	−30.06***	−30.06***
BIST TEKİRDAĞ	6.19	6.19	6.19	6.19	6.19	6.19	6.19
Pos: Neg	02:10						

Notes: All values are expressed in percentage (%). The symbols of ***, **, and * denote significance levels at %1, %5, and %10.

Tab. 6: Cumulative Average Abnormal Returns for Event Window (–20; 20)

CAAR [–20,20]							
City Indice	CAAR	Patell Z	Adj. Patell Z	BMP	KP	Wilcoxon	Grank
BIST ADANA	4.71	4.71	4.71	4.71	4.71	4.71	4.71
BIST ANKARA	18.18***	18.18***	18.18***	18.18***	18.18***	18.18***	18.18***
BIST ANTALYA	–25.96**	–25.96**	–25.96**	–25.96**	–25.96**	–25.96**	–25.96**
BIST BALIKESİR	–24.56**	–24.56**	–24.56**	–24.56**	–24.56**	–24.56**	–24.56**
BIST BURSA	–14.71*	–14.71*	–14.71*	–14.71*	–14.71*	–14.71*	–14.71*
BIST DENİZLİ	–13.22	–13.22	–13.22	–13.22	–13.22	–13.22	–13.22
BIST İSTANBUL	–2.72 %	–2.72 %	–2.72 %	–2.72 %	–2.72 %	–2.72 %	–2.72 %
BIST İZMİR	–6.09	–6.09	–6.09	–6.09	–6.09	–6.09	–6.09
BIST KAYSERİ	–22.80 **	–22.80 **	–22.80 **	–22.80 **	–22.80 **	–22.80 **	–22.80 **
BIST KOCAELİ	–0.68	–0.01	–0.01	–0.01	–0.01	–0.01	–0.01
BIST KONYA	–0.16	–0.16	–0.16	–0.16	–0.16	–0.16	–0.16
BIST TEKİRDAĞ	–0.13	–0.13	–0.13	–0.13	–0.13	–0.13	–0.13
Pos: Neg	02:10						

Notes: All values are expressed in percentage (%). The symbols of ***, **, and * denote significance levels at %1, %5, and %10.

Tab. 6 presents the CAARs for the event window (−20; 20). BIST Adana and BIST Ankara indices have positive returns however BIST Adana is insignificant. In general, the CAARs are not significant at 0.01 level but only BIST Ankara. BIST Antalya has still a negative abnormal return at 0.05 significance level. BIST Balıkesir follows, and it is significant at 0.05 level as well. The findings reveal the impact of 11 March, 2020 on city indices, and causes of the dramatical losses in analyzed event windows. Contrarily, the date of 10 April, 2020 when the curfews are imposed to 31 cities has found no impact on indices. The results for 10 April, 2020 are given in Appendix.

Conclusion

The COVID-19 pandemic was the unexpected event in recent days. By the outbreak of COVID-19, the markets are dramatically affected. The pandemic has arisen the uncertainty, trigger volatility and paralyzed both investors and global financial markets. Since the market indices are the indicator of financial performances, in this research, it has been attempted to measure the impact of COVID-19 on BIST City Indices in Turkey by employing Event Study approach. For that aim, when the COVID-19 pandemic announced on 11 March, 2020, and the curfew imposed to 31 cities on 10 March of 2020 are accepted as the event dates. The event windows are chosen in line with the literature as (−1; 1), (−2; 2), (−5; 5), (−10; 10), and (−20; 20). The market model is employed, and the CAARs are calculated. The findings present the severely loss in returns in the event windows of 11 March, 2020. Contrarily, the curfew announcement has no impact on city indices. The event window (−10; 10) exhibits the highest losses in city indices, specifically for BIST Antalya. It records 52.15 % loss. Besides, BIST Denizli, BIST Konya, BIST Balıkesir, BIST Adana, BIST Bursa follow with negative CAARs as 36.88 %, 30.06 %, 26.49 %, 23.16 % and 22.40 %, respectively. The CAARs are also high in the event window (−5; 5) and following (−2; 2) event window. In general, BIST Antalya indice provides the lowest CAAR values. It may stem from the impact of COVID-19 on tourism sector. Since BIST Antalya indice consists of the companies those run in tourism sector so the deep impact of sector dominates the indice return. It also reveals how Antalya city's economic situation is affected from the pandemic.

In general, it is observed that the economic situation of cities is seriously affected from pandemic. To be more specific, BIST Antalya and BIST Denizli are the mostly affected cities. As can be seen clearly, almost none of the event window, these indices turn to positive. As mentioned on daily news, the effects of the pandemic spread all over the city of Antalya in Turkey, and cause the tourism revenue to fall. The economic facilities slow down across the city from recreation agencies to hotels and restaurants. Just before the outbreak occurred, the hospitality industry had been preparing for the opening of season by March, and all those preparations required investment on human and tangible assets. That's why the slowdown in city folds the economic loses. The negative effects of pandemic have been felt deeply by residents, investors, the owners of hotels, restaurant, travel agencies, and beyond.

Declaration of Conflicting Interests

The authors declared no potential conflicts of interest with respect to the research, authorship and/or publication of this article.

Funding

The authors received no financial support for conducting research, authorship and/or publication of this article.

Acknowledgements

I thank Dr. Fausto Pacicco for the support in running Stata "estudy" command and to have conducted the statistical analysis of the study.

Appendix: The Results of Event Dated on 10 April, 2020

Tab. 7: Cumulative Average Abnormal Returns for Event Window (−1; 1)

CAAR [−1,1]								
City Indice	CAAR	Patell Z	Adj. Patell Z	BMP	KP	Wilcoxon	Grank
BIST ADANA	−4.59	−11.45***	−11.45***	−11.45***	−11.45***	−11.45***	−11.45***
BIST ANKARA	−2.49	−4.14***	−4.14***	−4.14***	−4.14***	−4.14***	−4.14***
BIST ANTALYA	3.34	−20.85***	−20.85***	−20.85***	−20.85***	−20.85***	−20.85***
BIST BALIKESİR	5.39*	−12.22***	−12.22***	−12.22***	−12.22***	−12.22***	−12.22***
BIST BURSA	−0.08	−3.86*	−3.86*	−3.86*	−3.86*	−3.86*	−3.86*
BIST DENİZLİ	0.03	−21.19***	−21.19***	−21.19***	−21.19***	−21.19***	−21.19***
BIST İSTANBUL	1.22	−4.28***	−4.28***	−4.28***	−4.28***	−4.28***	−4.28***
BIST İZMİR	0.32	−8.98***	−8.98***	−8.98***	−8.98***	−8.98***	−8.98***
BIST KAYSERİ	−1.83	−11.52***	−11.52***	−11.52***	−11.52***	−11.52***	−11.52***
BIST KOCAELİ	−2.71	−4.75**	−4.75**	−4.75**	−4.75**	−4.75**	−4.75**
BIST KONYA	−0.8	−19.52***	−19.52***	−19.52***	−19.52***	−19.52***	−19.52***
BIST TEKİRDAĞ	1.88	7.90**	7.90**	7.90**	7.90**	7.90**	7.90**
Pos: Neg	07:05						

Notes: All values are expressed in percentage (%). The symbols of ***, **, and * denote significance levels at %1, %5, and %10.

Tab. 8: Cumulative Average Abnormal Returns for Event Window (−2; 2)

CAAR [−2,2]							
City Indice	CAAR	Patell Z	Adj. Patell Z	BMP	KP	Wilcoxon	Grank
BIST ADANA	−4.73	−11.62***	−11.62***	−11.62***	−11.62***	−11.62***	−11.62***
BIST ANKARA	2.61	−4.72**	−4.72**	−4.72**	−4.72**	−4.72**	−4.72**
BIST ANTALYA	0.46	−25.01***	−25.01***	−25.01***	−25.01***	−25.01***	−25.01***
BIST BALIKESİR	5.66	−15.45***	−15.45***	−15.45***	−15.45***	−15.45***	−15.45***
BIST BURSA	−0.51	−9.85***	−9.85***	−9.85***	−9.85***	−9.85***	−9.85***
BIST DENİZLİ	4.00	−23.75***	−23.75***	−23.75***	−23.75***	−23.75***	−23.75***
BIST İSTANBUL	2.36*	−3.72***	−3.72***	−3.72***	−3.72***	−3.72***	−3.72***
BIST İZMİR	2.17	−10.34***	−10.34***	−10.34***	−10.34***	−10.34***	−10.34***
BIST KAYSERİ	−0.88	−13.96***	−13.96***	−13.96***	−13.96***	−13.96***	−13.96***
BIST KOCAELİ	−2.44	−5.64**	−5.64**	−5.64**	−5.64**	−5.64**	−5.64**
BIST KONYA	0.7	−20.99***	−20.99***	−20.99***	−20.99***	−20.99***	−20.99***
BIST TEKİRDAĞ	−1.41	−2.88	−2.88	−2.88	−2.88	−2.88	−2.88
Pos: Neg	07:05						

Notes: All values are expressed in percentage (%). The symbols of ***, **, and * denote significance levels at %1, %5, and %10.

Tab. 9: Tab. 3: Cumulative Average Abnormal Returns for Event Window (−5; 5)

CAAR [−5,5]							
City Indice	CAAR	Patell Z	Adj. Patell Z	BMP	KP	Wilcoxon	Grank
BIST ADANA	3.58	−23.93***	−23.93***	−23.93***	−23.93***	−23.93***	−23.93***
BIST ANKARA	3.32	−8.35***	−8.35***	−8.35***	−8.35***	−8.35***	−8.35***
BIST ANTALYA	7.31	−39.19***	−39.19***	−39.19***	−39.19***	−39.19***	−39.19***
BIST BALIKESİR	4.57	−21.14***	−21.14***	−21.14***	−21.14***	−21.14***	−21.14***
BIST BURSA	8.92**	−23.29***	−23.29***	−23.29***	−23.29***	−23.29***	−23.29***
BIST DENİZLİ	8.54	−36.88***	−36.88***	−36.88***	−36.88***	−36.88***	−36.88***
BIST İSTANBUL	1.87	−6.75***	−6.75***	−6.75***	−6.75***	−6.75***	−6.75***
BIST İZMİR	2.83	−21.01***	−21.01***	−21.01***	−21.01***	−21.01***	−21.01***
BIST KAYSERİ	−2.76	−21.00***	−21.00***	−21.00***	−21.00***	−21.00***	−21.00***
BIST KOCAELİ	2.25	−21.46***	−21.46***	−21.46***	−21.46***	−21.46***	−21.46***
BIST KONYA	8.77	−36.11***	−36.11***	−36.11***	−36.11***	−36.11***	−36.11***
BIST TEKİRDAĞ	−1.61	−21.44***	−21.44***	−21.44***	−21.44***	−21.44***	−21.44***
Pos: Neg	10:02						

Notes: All values are expressed in percentage (%). The symbols of ***, **, and * denote significance levels at %1, %5, and %10.

Tab. 10: Cumulative Average Abnormal Returns for Event Window (−10; 10)

CAAR [−10,10]								
City Indice	CAAR	Patell Z	Adj. Patell Z	BMP	KP	Wilcoxon	Grank
BIST ADANA	17.33**	−23.38***	−23.38***	−23.38***	−23.38***	−23.38***	−23.38***
BIST ANKARA	3.94	0.71	0.71	0.71	0.71	0.71	0.71
BIST ANTALYA	29.14***	−52.49***	−52.49***	−52.49***	−52.49***	−52.49***	−52.49***
BIST BALIKESİR	25.79***	−27.24***	−27.24***	−27.24***	−27.24***	−27.24***	−27.24***
BIST BURSA	15.20***	−22.46***	−22.46***	−22.46***	−22.46***	−22.46***	−22.46***
BIST DENİZLİ	27.64***	−36.81***	−36.81***	−36.81***	−36.81***	−36.81***	−36.81***
BIST İSTANBUL	1.54	−4.92**	−4.92**	−4.92**	−4.92**	−4.92**	−4.92**
BIST İZMİR	9.71**	−13.78***	−13.78***	−13.78***	−13.78***	−13.78***	−13.78***
BIST KAYSERİ	14.09*	−22.69***	−22.69***	−22.69***	−22.69***	−22.69***	−22.69***
BIST KOCAELİ	10.30*	−15.22**	−15.22**	−15.22**	−15.22**	−15.22**	−15.22**
BIST KONYA	21.96**	−30.36***	−30.36***	−30.36***	−30.36***	−30.36***	−30.36***
BIST TEKİRDAĞ	15.28	5.70	5.70	5.70	5.70	5.70	5.70
Pos: Neg	12:00						

Notes: All values are expressed in percentage (%). The symbols of ***, **, and * denote significance levels at %1, %5, and %10.

Tab. 11: Cumulative Average Abnormal Returns for Event Window (−20; 20)

CAAR [−20,20]							
City Indice	CAAR	Patell Z	Adj. Patell Z	BMP	KP	Wilcoxon	Grank
BIST ADANA	26.99**	4.43	4.43	4.43	4.43	4.43	4.43
BIST ANKARA	10.82	17.73***	17.73***	17.73***	17.73***	17.73***	17.73***
BIST ANTALYA	29.29**	−26.49**	−26.49**	−26.49**	−26.49**	−26.49**	−26.49**
BIST BALIKESİR	35.53***	−26.12**	−26.12**	−26.12**	−26.12**	−26.12**	−26.12**
BIST BURSA	1.20	−14.64*	−14.64*	−14.64*	−14.64*	−14.64*	−14.64*
BIST DENİZLİ	36.04**	−13.39	−13.39	−13.39	−13.39	−13.39	−13.39
BIST İSTANBUL	7.00**	−2.92	−2.92	−2.92	−2.92	−2.92	−2.92
BIST İZMİR	17.79***	−8.39	−8.39	−8.39	−8.39	−8.39	−8.39
BIST KAYSERİ	21.72**	−19.65*	−19.65*	−19.65*	−19.65*	−19.65*	−19.65*
BIST KOCAELİ	11.13	−0.02	−0.02	−0.02	−0.02	−0.02	−0.02
BIST KONYA	38.23***	−16.28	−16.28	−16.28	−16.28	−16.28	−16.28
BIST TEKİRDAĞ	34.93**	−13.71	−13.71	−13.71	−13.71	−13.71	−13.71
Pos: Neg	12:00						

Notes: All values are expressed in percentage (%). The symbols of '***', '**', and '*' denote significance levels at %1, %5, and %10.

Bibliography

Açikgöz, Ö. and Asli G. (2020), The Early Impact of the COVID-19 Pandemic on The Global and Turkish Economy, *Turkish Journal of Medical Sciences*, 50 (SI-1), 520–526.

Alber, N. (2020), The Effect of Coronavirus Spread on Stock Markets: The Case of The Worst 6 Countries, *Available at SSRN 3578080*.

Albulescu, C. (2020a), Coronavirus and Financial Volatility: 40 Days of Fasting and Fear, *arXiv preprint arXiv:2003.04005*.

Albulescu, C. (2020b), Coronavirus and Oil Price Crash, *Available at SSRN 3553452*.

Al-Awadhi, A. M., Alsaifi, K., Al-Awadhi, A., and Alhammadi, S. (2020), Death and Contagious Infectious Diseases: Impact of the COVID-19 Virus on Stock Market Returns, *Journal of Behavioral and Experimental Finance*, 27, pp. 100–326.

Armitage, S. (1995), Event Study Methods and Evidence on Their Performance, *Journal of Economic Surveys*, 9 (1), pp. 25–52.

Baker, S. R., Bloom, N., Davis, S. J., Kost, K. J., Sammon, M. C., and Viratyosin, T. (2020), *The Unprecedented Stock Market Impact of COVID-19*, No. w26945. National Bureau of Economic Research.

Bayramoğlu, M. F. and Pekkaya, M. (2010), İMKB Tarafından Hesaplanan Endekslerde Yeni Gelişmeler ve İMKB Şehir Endeksleri, *Muhasebe ve Finansman Dergisi*, 45, pp. 200–215.

Beck, T. (2020), Finance in the Times of Coronavirus, *Economics in the Time of COVID-19*, pp. 73–76.

Borsa İstanbul (BIST), "City Indices", https://www.borsaistanbul.com/en/sayfa/2233/city-indices (Access Date: 25.08.2020).

Boehmer, E., Masumeci, J., and Poulsen, A. B. (1991), Event-Study Methodology Under Conditions of Event-Induced Variance, *Journal of Financial Economics*, 30 (2), pp. 253–272.

Chowdhury, E. K. and Abedin, M. Z. (2020), COVID-19 Effects on The US Stock Index Returns: An Event Study Approach, *Available at SSRN 3611683*.

Fama, E. F. (1970), Efficient Capital Markets: A Review of Theory and Empirical Work, *Journal of Finance*, 25 (2), pp. 385–417.

Fama, E. F. (1991), Efficient Capital Markets: II, *The Journal of Finance*, 46 (5), pp. 1575–1617.

Feng, J., Bao, Y., Wang, Y., Meng, S., Xia, J., and Zhang, T. (2020), Coronavirus vs Market: Investment Opportunities Lies Underneath the Epidemic. *Available at SSRN 3563059*.

Göker, İ. E. K., Eren, B. S., and Karaca, S. S. (2020), The Impact of the COVID-19 (Coronavirus) on The Borsa Istanbul Sector Index Returns: An Event Study, *Gaziantep Üniversitesi Sosyal Bilimler Dergisi*, 19 (COVID-19 Special Issue), pp. 14–41.

Kılıç, Y. (2020), Borsa İstanbul'da COVID-19 (Koronavirüs) Etkisi, *JOEEP: Journal of Emerging Economies and Policy*, 5 (1), pp. 66–77.

Kolari, J. W. and Pynnönen, S. (2010), Event Study Testing with Cross-Sectional Correlation of Abnormal Returns, *The Review of Financial Studies*, 23 (11), pp. 3996–4025.

Luo, S. and Tsang, K. P. (2020), How Much of China and World GDP Has The Coronavirus Reduced?, *Available at SSRN 3543760*.

Makridis, C. and Hartley, J. (2020), The Cost Of COVID-19: A Rough Estimate of the 2020 US GDP Impact, *Special Edition Policy Brief*.

Özkan, N. and Ulaş, Ü. (2021), Bölgesel Covid-19 Vaka Sayilari, Altin Fiyatlari, Euro ve BIST Şehir Endeksleri Arasindaki İlişki: Bir ARDL Sinir Testi Yaklaşımı, *Ekonomi Politika ve Finans Araştırmaları Dergisi*, 6 (1), pp. 240–253.

Patell, J. M. (1976), Corporate Forecasts of Earnings Per Share and Stock Price Behavior: Empirical Test, *Journal of Accounting Research*, 14 (2), pp. 246–276.

Sansa, N. A. (2020), The Impact of the COVID-19 on the Financial Markets: Evidence from China and USA, *Electronic Research Journal of Social Sciences and Humanities*, 2, pp. 29–39.

Sarı, S. S. and Kartal, T. (2020), COVID-19 Salgininin Altin Fiyatlari, Petrol Fiyatlari ve VIX Endeksi Ile Arasindaki Ilişki, *Erzincan Üniversitesi Sosyal Bilimler Enstitüsü Dergisi*, 13 (1), pp. 93–109.

Şenol, Z. and Zeren, F. (2020), Coronavirus (COVID-19) and Stock Markets: The Effects of the Pandemic on the Global Economy, *Avrasya Sosyal ve Ekonomi Araştırmaları Dergisi*, 7 (4), pp. 1–16.

World Health Organization (WHO). WHO Coronavirus (COVID-19) Dashboard. https://COVID19.who.int/table, (28.05.2020).

Yan, C. (2020), COVID-19 Outbreak and stock prices: Evidence from China, *Available at SSRN 3574374*.

Zeren, F. and Hızarcı, A. (2020), The Impact of COVID-19 Coronavirus on Stock Markets: Evidence from Selected Countries, *Muhasebe ve Finans İncelemeleri Dergisi*, 3 (1), pp. 78–84.

Zhang, D., Hu, M., and Ji, Q. (2020), Financial Markets under the Global Pandemic of COVID-19, *Finance Research Letters*, 36, 101528.

Alper ATEŞ and Ömür Hakan KUZU

Tourism Higher Education Management and Policies in Times of Crisis

1 Introduction

Despite the risks of disappearing in the future, the tourism sector, which has unveiled and continues to unveil many occupational groups (Ayres, 2006; Mahdawi, 2017), is confronted with the problem of struggling with crises (Jamal & Budke, 2020) in addition to the paradoxical situation within. It is inevitable that tourism education, appraised in the category of vocational education (Cooper & Shepherd, 1997) due to its close relations with the sector, will also be hit by these crises. As in the examples of economic or political crises in the past, it is discerned that global epidemics such as Covid-19 cause serious losses (Çalıkoğlu & Gümüş, 2020; Salmi, 2020) in terms of higher education level learning outcomes, graduate qualifications, youth employment, inequalities and international integrations.

Apart from the fact that the Covid-19 epidemic offers positive possibilities in the most fundamental aspects such as digital skills and continuous learning (IAU, 2020; Soroya, Rehman, Abbas, Mirza, Mahmood, & Aboidullah, 2020); it is witnessed that there exist several problematic areas such as face-to-face education requirements in the new conditions, vocational and applied training difficulties, and internship as well as the increase in public expenditures in higher education, international student adaptation, loss of learning time, school dropouts, headaches in distance education accessibility, digital unpreparedness of academic members, the uncertainty of teaching times (OECD, 2020).

The fact that these uncertainties, which have emerged even more intensely during the crisis period, happen to be in a period when higher education is in a transformation (Matei, 2021), causes the concepts of higher education and crisis to be mentioned together. Because, since higher education is a modern age institution, the global epidemic endured during its adaptation to the post-modern age has brought a new set of obstacles for higher education institutions as in every crisis, and has caused these struggles to be viewed as a door of opportunity that should have been estimated long ago (Erkut, 2020). Thus, in fact, universities had to reveal good/bad pro-type of the practices required by the university understanding of the 21st century, albeit by force, as in the example of distance education (Hodges, Moore, Lockee, Trust, & Bond, 2020).

What has been experienced in higher education institutions in the last 30 years, in a sense, betokens the existence of a complex circumstance (Salmi, 2002) that arises with the interaction of opportunities and threats appearing as a result of competition, restructuring and a series of concepts/models. The fact that competition has grown to be a mechanism accommodating both regional and international cooperation within the understanding of neoliberalism, the emergence of new structures and models as well as the flexibility and diversification of learning models, and the spread of new management models such as strategic quality management and governance raise the pressure on higher education institutions. Under this pressure, it is predicted that universities will face more chaotic and unpredictable types (Almaraz-Menendez, Maz-Machado, & Lopez-Estabean, 2016) that offer opportunities during and after the global epidemic.

Although empirical studies on the post-Covid-19 situation of higher education reveal many results, it is seen that two critical issues rise to prominence. The first of these is the emphasis on a more sustainable education (Navarro-Espinosa, Vaquero-Abellán, Perea-Moreno, Pedrós-Pérez, Aparicio-Martínez, & Martínez-Jiménez, 2021) and the second, a more sectoral education (THE, 2020). The former is a more technical/educational discussion ranging from the human to the technological aspect while the latter brings along a more specific/industrial dispute that varies from social to commercial.

In this general discussion environment, debates on sustainability models have become extensive in the literature -especially for post-Covid-19- due to both the technological/digital transformation tendency of tourism higher education and its sectoral collaborative nature. These presented models have essentially been designed since the early 2000s for changing phenomena in the sector in the context of neo-liberal and post-capitalist socio-economic structures (Sheldon, Fesenmaier, Woeber, Cooper, & Antonioli, 2008; Tribe, 2001). It is such a coincidence that these determinations are presented as a basis for prescriptions (Tiwari, Séraphin, & Chowdhary, 2020), which are also presented as detections for Covid-19.

In this study, firstly, the change in universities in the context of crises in higher education is briefly mentioned, and then the subject of transformation in tourism higher education is reviewed. In the conclusion part, the overlapping elements of this transformation with the Covid-19 epidemic were analyzed; what the suggestions in the literature in terms of tourism higher education management and policies mean for post-Covid-19, and buzzwords of higher education such as stakeholder relations, quality management, sustainability, governance, entrepreneurship, innovation and digitalization are evaluated in terms of the transformation.

2 Crisis in Higher Education: The End of the Golden Age

Higher education, as an institution of the modern age, has been in a fast transformation that can be denominated as a crisis for a long time. As for crisis; close concepts such as problem, risk, uncertainty, conflict, stress, tension, chaos, and phenomena such as emergency and catastrophe (Milasinovic & Kesetovic, 2008) are definitions that summarize the circumstance, it, thus, would not be wrong to say that universities have been in crisis for a long time. Although it has reached the stage of losing its own potential and perfection (especially in developing countries, not for the pursuit of science and truth, but as a national development agency for the benefit of certain classes) (Timur, 2000), it is asserted that these crises will eventually open the doors of opportunity (Salmi, 2002).

Higher education institutions are in separately looking-extraordinary advancements in developed and developing countries separately (Timur, 2000). The deficiencies most influenced by the institutional developments in the universities of developed countries are the increasing economic/financial problems (Altbach, 2011); thus, more collective/social dilemmas have started to emerge in developing countries (GPSS, 2020). This reveals the regional differentiation in the historical development of higher education, but on the other hand, it does not prevent countries from being assessed in the same category (Altbach, 2015). Many universities are evaluated together with measurable criteria to achieve a more international and collaborative structure, which makes similarity and emulation models even more common (Üsdiken, Divarcı Çakmaklı, & Topaler, 2017). Moreover, despite this, many universities and countries wish to highlight differentiation as a model (Reimer & Jacob, 2010).

Transformation in higher education has been about what societies and countries expect from the university, and the change in these expectations put a strain on the university (Timur, 2000). In these periods, which can be considered as a transition of three and maybe four generations (Etzkowitz, & Leydesdorff, 2000; Lukovics & Zuti, 2015; Wissema, 2009), universities, as organizations with different priorities in terms of mission, have always existed in a double interactive transformation aimed at shaping societies and being affected by social developments (Zaglul, Sherrard, & Juma, 2006).

Universities, which proceeded by interpolating the research mission to the education mission and then the social service mission that sanctifies the research mission for social purposes, eventually reached an important position in the socio-economic structure between the state and the industry (Etzkowitz, 2013). This position has led universities to fight for international competition as well as supporting national development, which is the necessity of the modern age (Qiang, 2003). With the articulation of technological movement/digital transformation into this competition (Duderstadt, 2001), universities have become diverse institutions (Torres & Rhoades, 2006).

For this very reason, stress/conflict in the nature of crises has been the basic concepts that characterize the historical expansion of higher education institutions. In Tab. 1, the components of change in higher education in the literature are shown with the elements they conflict/opposite. Tab. 1 has also been believed as an indication that the establishment of the relationship between the new concepts of universities in change and the context has always been a series of crises in this change, yet it has the potential to accelerate forward despite the crises. This, however, does not mean there are no grave criticisms on the progressive view, and it is also asserted that universities are the institutions most impressed by globalization and crises today (Kurul Tural, 2004).

Tab. 1: Conflicts/Contradictions in the Change Paradigms of Higher Education

Change Paradigms	Education	Research	Social Service	Conflict
Globalization	Transnational university structures	International competition and cooperation	International mobility	Nation-state
Post-modern competition	Quality assurance and accreditation	Ranking institutions and indexed publication	Differentiation/ diversity in higher education supply	Academic freedom
Governance/ Managerialism	Financial diversity	Entrepreneurial/ innovative research university	Accountability and transparency in higher education	Institutional autonomy
Digitalization	Distance/ online/open learning	Multi-disciplinarity	Lifelong learning	"Educere" Science-Truth
Sustainability	Flexible models based on learning by doing and student centeredness	Teaching-reserach nexus	Industry/ stakeholder engagement in higher education	Academic capitalism

Source: Adapted by authors.

As can be seen in Tab. 1, with the transition of higher education from the modern university understanding to the post-modern university perception; the phenomena of globalization, competition, digitalization, managerialism and sustainability influence universities in the form of cause and effect. The effects of these phenomena on the three basic missions of universities also cause new formations in context, yet these developments also generate problematic areas on the nation-state, academic freedom, and scientific knowledge, the real purpose of education, institutional autonomy and social differences. So and so, it is an inevitable result that globalization threatens the existence of the nation-state (Habermas, 2018) and destroys the structure of universities, which are modern state institutions. There exists a post-modern understanding of competition that carries this point forward; the quality assurance system (Morley, 2003) and ranking institutions (Pusser & Marginson, 2013) narrow the freedom of universities/academics in the context of education and research activities (Ercan, 2011; Kalfa & Taksa, 2016). The fact that higher education gradually reveals the profile of students and employees who have become commercial parties in the university (Kandiko, 2010) with the effect of neo-liberal management (Özsoy, 2011) also leads the phenomenon of academic capitalism (Slaughter & Laslie, 1997) to become well established in universities. When the unresolvable contradictions occur even in the most basic developments such as distance learning as a result of digital transformation (Lugosi & Jameson, 2017) to all these, the issue of how the university will overcome the transformation pains/crises becomes more complicated.

3 Effects of Covid-19 on Higher Education: Normalization of Crises

With the outbreak of the Covid-19 epidemic, higher education institutions tried to eliminate the pressure on them by going into a general state of alarm around the world. The higher education system, already in pain in discerning the transformations required by the age, has passed into a state of questioning the status quo with the epidemic crisis. Views have increased in that this questioning will necessitate radical changes by removing the rigid/resisting structures/understandings in higher education systems. It is declared that some new elements should be added in the education, research and social service missions, the main missions of higher education institutions during Covid-19. It is also asserted that the opportunities it offers for internationalization (globalization) and governance mechanisms should be evaluated well, otherwise it will not be able to provide a solution to empty campuses, dysfunctional research laboratories and the system with frozen international mobility (Recio & Colella, 2020).

Universities had to expand distance education mechanisms very quickly with the global epidemic. In a period when there are still disputes on the distance education system (Çalıkoğlu & Gümüş, 2020), such a forced transition has even deepened the problems, but it has also brought new experiences to universities in dealing with problems and taking care of opportunities, as in every crisis period (Adedoyin & Soykan, 2020). However, as well as the difficulties that can be summarized as distance/online learning, technological inadequacies (difficulty in accessing technology), the inadequacy of socio-economic conditions in accessing the internet, unwanted interventions in the home environment, digital competence, measurement and evaluation problems, and problems in areas requiring application during Covid-19; there exist opportunities such as the emergence of new (innovative) research areas, especially Covid-19 and distance learning, the increase of new (innovative) technologies, and the spread of socio-economic interventions and supports for students (Adedoyin & Soykan, 2020). Besides, the Covid-19 outbreak has also led to the emergence of a double-interactive awareness as a way of promoting and recognizing digital learning. This awareness process has conferred the necessity of obtaining the unconditional adoption of online learning technologies (Todorova & Bjorn-Andersen, 2011) before the disaster/crisis begins, as a lesson to universities during the emergency remote teaching period (Adedoyin & Soykan, 2020).

Fig. 1 manifests the strengths-weaknesses of the distance/online learning system and the findings of opportunity-threat/challenge analysis according to the crisis periods in the world since the 2000s. According to this analysis, it is understood that the system has technical and individual struggles and problems as well as its strengths in the form of time-place flexibility, fast and widespread access. On the other hand, while the system offers opportunities in terms of age, program, skill and innovation, it also has some disadvantages in terms of social inequality in accessing computer technologies, quality of education, digital literacy and cost. Nevertheless, despite everything, it is underlined that online learning is also a panacea during the Covid-19 crisis (Dhawan, 2020). Hence, it is said that incentives for students and academics to use different online tools and for universities to promote digital teaching practices should continue after the pandemic (Pokhrel & Chhetri, 2021).

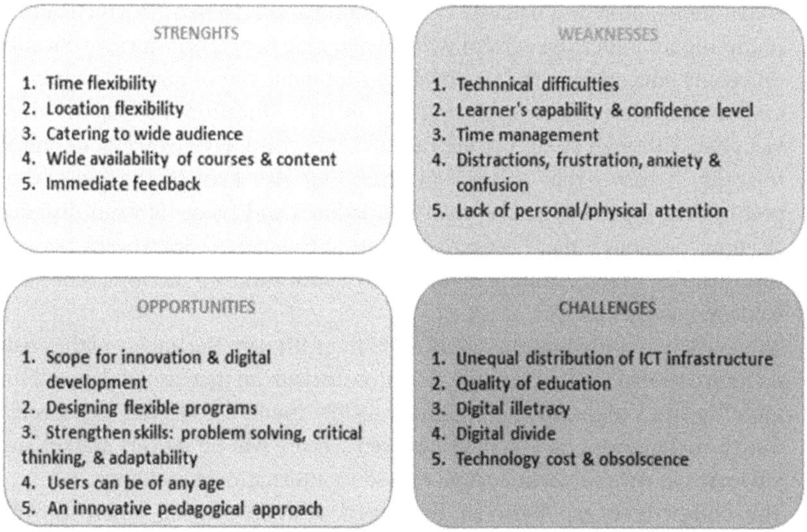

STRENGHTS	WEAKNESSES
1. Time flexibility 2. Location flexibility 3. Catering to wide audience 4. Wide availability of courses & content 5. Immediate feedback	1. Technnical difficulties 2. Learner's capability & confidence level 3. Time management 4. Distractions, frustration, anxiety & confusion 5. Lack of personal/physical attention

OPPORTUNITIES	CHALLENGES
1. Scope for innovation & digital development 2. Designing flexible programs 3. Strengthen skills: problem solving, critical thinking, & adaptability 4. Users can be of any age 5. An innovative pedagogical approach	1. Unequal distribution of ICT infrastructure 2. Quality of education 3. Digital illetracy 4. Digital divide 5. Technology cost & obsolscence

Fig. 1: SWOT/SWOC Analysis for Online Learning in Times of Crisis
Source: Dhawan, 2020: 10.

The possible situations/crises – with the example of American universities – that are expected to be experienced after the crisis with Covid-19 of contemporary higher education depending on globalization in terms of not only student mobility and internationalization initiatives but also collaborative research and increasingly global information networks and other aspects can be summarized as follows (Altbach & de Wit, 2020):

1. The fittest will survive: According to this prediction, strong research universities will continue to obtain more interest, but private universities with less funding, as in the case of America, will face the danger of closure.
2. Research topics will be monopolized: In terms of research, medicine and life sciences will become more popular, and employment and financial problems will arise for research/researchers in other areas.
3. There will be problems in public finance and employment: There will be employment problems in universities and uncertainties in the admission of students will continue. It seems possible that public allocations for higher education will decrease in the future due to the large expenditures of countries aimed at stabilizing the economies during the crisis.

4. Social inequalities and equality of opportunity will deepen: The inequalities in the higher education system will deepen and people will turn to cheaper university education in the face of the unstoppable rise of mass education.
5. Online education will continue, but face-to-face education will become more valuable: Although all universities around the world have switched to online learning, it shows that online education will also exist in the future, but problematic areas will continue due to failures and inequalities in distance learning. Although there is no revolution in this regard, improvements will continue. However, traditional face-to-face education will become more valuable/privileged.
6. There will be a shift in international mobility: Although the impact of the crisis on international student mobility remains unclear, an increase in demand for mobility from high-income countries to low-income countries is expected due to cost increase and security concerns. There will be serious market and job losses in America and Europe related to international exchange.
7. The competitive gap between developed and developing countries will increase: The effects of the crisis on higher education will be significant and mostly negative, and gaps and inequalities between students, institutions and countries will grow. Global disparities between universities in developed and developing countries will become more significant as universities in the world's poorest parts of the world are likely to be more severely affected by the crisis.

While similar scenarios are envisioned for Australian universities, especially regarding the funding of research universities and the drop in international mobility, universities may have to respond to the pressures of building programs to meet the expectations of prospective students, including short and non-credit courses (Croucher & Locke, 2020). After all these expectations, although it appears possible to continue on the road without reacting to higher education institutions or to react later, the main choice will be between transformation and optimizing options. Whichever option is applied, universities are expected to review their strategies, improve their core capabilities, adapt a target-operating model, and modernize their technologies in the coming days (KPMG, 2020). This will also call for an increase in societal participation and public service responsibility in terms of external stakeholders/external environment of higher education, and in terms of internal stakeholders/internal environment, which will lead to participatory governance models and approaches that guarantee student satisfaction and welfare, as in crisis management models (Recio & Colella, 2020).

4 Tourism Higher Education and Covid-19: The New Normal-Old Prescription

As a reflection of the technical, practical and liberating benefits of scientific positivist, interpretive method and critical theory in higher education; a curriculum that ignores conflicting values, is based on measurable facts (facts), gives importance to multiple understandings of the tourism world, is equidistant from values, and therefore less political is regarded as beneficial in tourism higher education- due to the comprehensiveness of tourism. It is asserted that the tourism higher education curriculum should be designed not only for the business world but also for all segments of society and the advancement of regional tourism. A curriculum based on all three approaches, as well as non-professional philosophical aspects such as the purpose and meaning of education, is believed to be a requirement of the age for tourism graduates (Tribe, 2001).

What is said about sustainability in today's tourism higher education is replicated after Covid-19. In addition to what was said at the beginning of the century, Tourism Education Futures Initiative (TEFI), an initiative with the widest participation in designing tourism higher education, has decided to consider administrative (technical), moral (political and ethical), human (social) and professional (practical) abilities in tourism higher education. In this model, while technical skills such as using real and virtual networks, sharing information, and environmental management skills are expected from graduates, they are also morally required to have cultural equipment that exhibits ethical behavior and integrates basic human values. It was stated that there should be learning outcomes such as flexibility, multitasking, critical thinking, innovation/entrepreneurship, communication with digital tools, multi-cultural competencies, risk management, and problem-solving skills from a professional perspective. (Sheldon, Fesenmaier, Woeber, Cooper, & Antonioli, 2008).

Due to the mutual interactional structures of tourism higher education and tourism sector, it has been inevitable to experience what happened in the tourism sector during the Covid-19 crisis also in tourism higher education, and problems have started to occur in the above-mentioned student gains. The necessity to reshape the industry for Covid-19 (such as adapting to altering consumer behavior, community orientation, and applications of artificial intelligence and robotics) has similarly applied to tourism higher education. This has resulted in tourism higher education, which is actually intertwined with the sector and has sectoral immanence, to establish more relations with the sector (Tiwari, Séraphin, & Chowdhary, 2020).

The fact that more digitalization in tourism higher education compels student-centered practices and the creation of more sustainable and environmentally and climate change-sensitive curriculum contents can be deemed as parallel developments with the sector. As in all areas of higher education, however, in parallel with sectoral developments in tourism higher education (Gretzel et al., 2020), it is inevitable that the scenarios, evolved after Covid-19, are based on computer technologies in terms of the harmony of education with the sector. In addition, it is expected that the state's support to tourism higher education institutions should be extended, just like the tourism sector, and that universities at the national and international level should engage in collaborative activities in this course (Tiwari, Séraphin, & Chowdhary, 2020).

In the study of Tiwari, Séraphin, and Chowdhary (2020), one of the first studies on the effects of Covid-19 on tourism higher education and post-Covid-19 scenarios, these scenarios are encoded as deterrent factors, insufficient student skills, and radical transformations in the education system and a more dynamic tourism curriculum. Accordingly, it is prognosticated that there will be unemployment, job losses, and job contractions in the sector because of travel restrictions, which therefore will induce a contraction in the demand for tourism higher education. It is also expressed that a more innovative and creative (project-oriented) education structure should be made to improve the current education, and tourism graduate candidates should be trained on issues such as health, safety, and risk management to raise awareness of current crisis problems. In addition to these pieces of training, it is underlined that the development of indispensable aspects of tourism education such as foreign language, computer use and communication skills should not be disregarded. The unconditional acceptance of the radical transformations in tourism higher education in terms of online education and the re-discovery of tourism higher education with a multi-disciplinary research focus, open stakeholder relations and sector participation in all educational processes, especially in the curriculum, is requested. In summary, it is pronounced that tourism higher education will become more robust after Covid-19, with cooperation with the sector in curriculum design and employment problems, state support in research financing, digital, innovative and vocational qualifications of students (Tiwari, Séraphin, & Chowdhary, 2020).

It would not be wrong to say that the prescriptions conferred above for the recovery after Covid-19 in tourism higher education and all areas of higher education in general do not grant new prescriptions in the new normal. The fact that the model in Fig. 2, which is proposed for crisis moments in developing countries and problematic higher education systems is similar to the literature, frames an important basis for this finding. Because the model proposes a vision based on four basic principles for societies influenced by emergencies and conflict: to realize the following visions; (1) protection (ensure the safety of students in times of crisis), (2) access (providing greater access to individuals affected by crisis), (3) inequality (focusing on the most vulnerable students in crisis), and (4) empowerment (empowering individuals and society with the equipment of higher education), it is aimed to eliminate the damage of the crises on higher education institutions and therefore on societies with more, better and faster implementation through (1) strengthening academic capacity, (2) encouraging more cooperation among stakeholders, (3) generating unused resources, (4) improving understanding and accountability, and (5) moving higher education to the center of the system by providing financial support with state and sector support (GPSS, 2020).

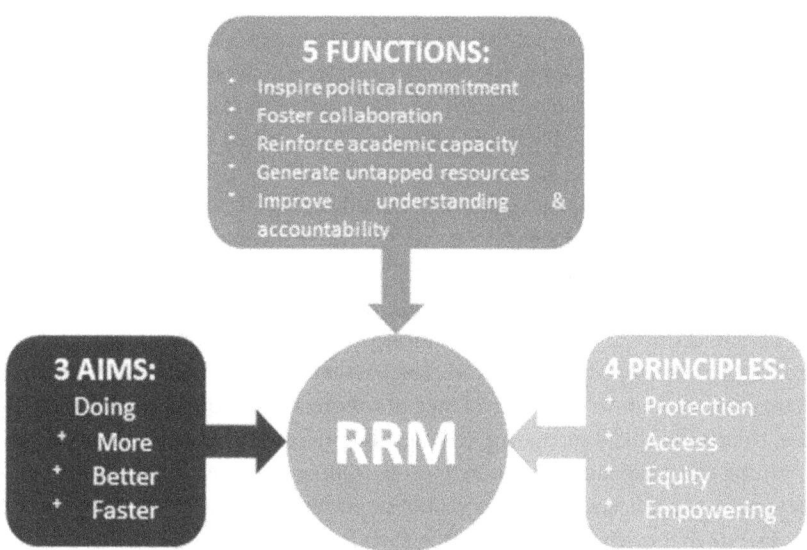

Fig. 2: Emergency Model in Higher Education
Source: GPSS, 2020: 64.

Conclusion

The Covid-19 outbreak has intensified the problems of learning outcomes/graduate qualifications, employment problems, inequalities (inequality of access to distance education), and disruption of internationalization in higher education institutions. However, it has made some enrichment to the acceleration of digitalization. To get out of this crisis; "globalizational purpose internationalization", "supra-competitive positioning", "governance" and "digitalization", which has long been presented as a recipe for higher education systems, has been presented as a starting point. Although it is asserted that universities expedite this transformation with their urgent digitalization efforts owing to the discourse that crises are a door of opportunity; there has even not been any significant progress made for the difficulty of accessing digital education for rich and poor countries, as well as for vulnerable and underprivileged students (Salmi, 2020), let alone becoming solved in Covid-19.

The fact that these arrangements in universities, which are expected to be restructured after the crisis, are also based on technology, increased the pressure on the financial resource problem on universities, and it was recommended to further develop relations with the state and the industry. The position of the state here is related to the longing for the social state, which came to the fore again during Covid-19. This can be recognized as the adaptation of the new public service understanding (Denhardt & Denhardt, 2000), which neo-liberalism based on using public resources for industry (capital) and has already shaped and taken one step further, to the governance understanding, that is, the presentation of the old recipe to the new normal.

Unlike other fields of higher education, despite the crises in the sector, the adaptability of tourism higher education, which is closely related to the tourism sector, has let this type of applied education-oriented higher education not to have problems in developed country examples. Furthermore, in parallel with the rapid changes in the industry, the prevalence of robotization and advanced technologies, the rise in social media applications, the increase in individual (boutique) solutions, the dominance of a more interdisciplinary perspective, and the implementation of applications such as virtual classrooms and internships are also on the agenda of tourism higher education (Seo & Kim, 2021). Notwithstanding the unstoppable ascent of distance education, an important tool in coping with difficulties, to avoid the problems experienced during Covid-19 in the next crises, as in other areas in tourism higher education; more supportive and compassionate, more student engaged-focused models in which active learning is effective are recommended especially in terms of student satisfaction (Bakeri, 2021). For example, according to one applied research in the USA on participation (Zhong, Busser, Shapoval, & Murphy, 2021), it is declared that students are satisfied with the participation mechanisms in terms of collaborative learning, student-faculty interaction, effective teaching practices, quality interaction and campus supportive environments.

At the same time, as in the study of Lei and So (2021) during the Covid-19 epidemic, the differences in satisfaction levels based on the perceived benefit/experience differences between students and academics about distance education make the subject of learning/teaching methods a more dynamic problem; successful examples of using new applications with a combination of distance learning/virtual practice for internships and laboratories improve the satisfaction levels and learning outcomes of tourism students (Orlowski, Mejia, Back, & Fridrich, 2021; Park & Jones, 2021), just like in other areas of higher education. It is also regarded as a sign that digital applications in higher education will become more widespread in times of crisis or after.

In conclusion, the inevitable technological revolution or digital transformation, and the fact that the expectations of the students/academics/employees and the industry-society at universities tend towards a diverse but essentially uniform and common achievement and the gaps in meeting expectations in times of crisis increase, justifies the words of E.O. Wilson's: "The real problem of humanity is the following: We have paleolithic emotions, medieval institutions and godlike technology. And it is terrifically dangerous, and it is now approaching a point of crisis overall." and shows that higher education institutions (universities) established in the Middle Ages but evolved in the modern age will always live in crises. In this sense, the preservation of the validity of the prescriptions of universities in the transition period to the global-digital age, even in the biggest crisis of the century such as Covid-19, reveals two results. After this or other crises, nothing will change for universities and the direction steered by global actors and capital will definitely be followed or, in Rumi's words, "new things will have to be said".

Bibliography

Adedoyin, O. B. and Soykan, E. (2020). Covid-19 pandemic and online learning: The challenges and opportunities. *Interactive Learning Environments*, 1–13. Retrieeved from https://www.tandfonline.com/doi/epub/10.1080/10494820.2020.1813180?needAccess=true (Access Date: December 7, 2021).

Almaraz-Menendez, F., Maz-Machado, A., and Lopez-Esteban, C. (2016). University strategy and digital transformation in higher education institutions: A documentary analysis. *International Journal of Advanced Research, 11*(4), 2284–2296.

Altbach, P. G. (2011). The past, present, and future of the research university. In P. G. Altbach and J. Salmi (Eds.), *The road to academic excellence the* making of world-class research universities (pp. 11–32). Washington DC: The World Bank.

Altbach, P. (2015). The dilemmas of ranking. *International Higher Education,* (42), 2–3.

Altbach, P. G. and de Wit, H. (2020). Postpandemic outlook for higher education is bleakest for the poorest. *International Higher Education, 102,* 3–5.

Ayres, H. J. (2006). Education and opportunity as influences on career development: Findings from a preliminary study in eastern Australian tourism. *Journal of Hospitality, Leisure, Sport and Tourism Education, 5*(1), 16–27.

Bakeri, M. A. (2021). Educational distancing: A mixed-methods study of student perceptions in the time of coronavirus. *Journal of Hospitality & Tourism Education, 33*(3), 207–221.

Çalıkoğlu, A. and Gümüş, S. (2020). Yükseköğretimin geleceği: Covid-19'un öğretim, araştırma ve uluslararasılaşma konularındaki etkileri. *Yükseköğretim Dergisi, 10*(3), 249–249.

Cooper, C. and Shepherd, R. (1997). The relationship between tourism education and the tourism industry: Implications for tourism education. *Tourism Recreation Research, 22*(1), 34–47.

Croucher, G. and Locke, W. (2020). *A post-coronavirus pandemic world: Some possible trends and their implications for Australian higher education.* Melbourne CSHE Discussion Paper. Retrieved from https://melbourne-cshe.unimelb.edu.au/__data/assets/pdf_file/0010/3371941/a-post-coronavirus-world-for-higher-education_final.pdf (Access Date: December 7, 2021).

Denhardt, R. B. and Denhardt, J. V. (2000). The new public service: Serving rather than steering. *Public Administration Review, 60*(6), 549–559.

Dhawan, S. (2020). Online Learning: A Panacea in the time of Covid-19 crisis. *Journal of Educational Technology Systems, 0*(0), 1–18.

Duderstadt, J. J. (2001). The future of the university in the digital age. *Proceedings of the American Philosophical Society, 145*(1), 54–72.

Ercan F. (2011). Ölçerim, ölçebilmek için biçimlendiririm, biçimlendiğinde standartlaşmışsındır, standartlaştığında yönetirim, ölçemediğimi yönetemem o zaman cezalandırır, dışlarım. F. Ercan ve S. Korkusuz Kurt (Ed.), *Metalaşma ve iktidarın baskısındaki üniver*site (ss. 361–388). İstanbul: SAV Yayınları.

Erkut, E. (2020). Covid-19 sonrası yükseköğretim. *Yükseköğretim Dergisi, 10*(2), 125–133.

Etzkowitz, H. (2013). Anatomy of the entrepreneurial university. *Social Science Information, 52*(3), 486–511.

Etzkowitz, H. and Leydesdorff, L. (2000). The dynamics of innovation: From national systems and "mode 2" to a triple helix of university-industry-government relations. *Research Policy, 29*(2), 109–123.

GPSS (2020). *Higher education in emergencies: On the road to 2030.* Global Platform for Syrian Students Report, 1–136. Retrieved from https://www.esu-online.org/wp-content/uploads/2020/04/Report-Higher-Education-on-the-road-to-2030.pdf (Access Date: December 7, 2021).

Gretzel, U., Fuchs, M., Baggio, R., Hoepken, W., Law, R., Neidhardt, J., ... Xiang, Z. (2020). E-tourism beyond covid-19: A call for transformative research. *Information Technology & Tourism, 22,* 187–203.

Habermas, J. (2018). *Küreselleşme ve milli devletlerin geleceği* (Çev. M. Beyaztaş). İstanbul: Yarın Yayıncılık.

Hodges, C., Moore, S., Lockee, B., Trust, T., and Bond, A. (2020). The difference between emergency remote teaching and online learning. *Educause Review.* Retrieved from https://er.educause.edu/articles/2020/3/the-difference-between-emergency-remote-teaching-and-online-learning (Access Date: December 7, 2021)..

IAU (2020). *Regional/national perspectives on the impact of covid-19 on higher education.* UNESCO House, Paris: International Association of Universities. Retrieved from https://www.iau-aiu.net/IMG/pdf/iau_covid-19_regional_perspectives_on_the_impact_of_covid-19_on_he_july_2020_.pdf (Access Date: December 7, 2021).

Jamal, T. and Budke, C. (2020). Tourism in a world with pandemics: Local-global responsibility and action. *Journal of Tourism Futures, 6*(2), 181–188.

Kalfa, S. and Taksa, L. (2016). Employability, managerialism, and performativity in higher education: A relational perspective. *Higher Education, 74*(4), 687–699.

Kandiko, C. B. (2010). Neoliberalism in higher education: A comparative approach. *International Journal of Arts and Sciences, 3*(14): 153–175.

KPMG (2020). *The future of higher education in a disruptive world.* KPMG International Limited Publishing. Retrieved from https://assets.kpmg/content/dam/kpmg/xx/pdf/2020/10/future-of-higher-education.pdf (Access Date: December 7, 2021).

Kurul Tural, N. (2004). *Küreselleşme ve üniversiteler.* Ankara: Kök Yayıncılık.

Lei, S. I. and So, A. S. I. (2021). Online teaching and learning experiences during the covid-19 pandemic- a comparison of teacher and student perceptions. *Journal of Hospitality & Tourism Education, 33*(3), 148–162.

Lugosi, P. and Jameson, S. (2017). Challenges in hospitality management education: Pespectives from the United Kingdom. *Journal of Hospitality and Tourism Management, 31*, 163–172.

Lukovics, M. and Zuti, B. (2015). New functions of universities in the XXI. century: "fourth generation" universities. *Transition Studies Review, 22*(2), 33–48.

Mahdawi, A. (2017). *What jobs will still be around in 20 years? Read this to prepare your future.* US news. The Guardian. Retrieved from https://www.theguardian.com/us-news/2017/jun/26/jobs-future-automation-robots-skills-creative-health (Access Date: December 7, 2021).

Matei, L. (2021). Covid-19 and "the crises in higher education". In S. Bergan, T. Gallagher, I. Harkavy, R. Munck and H. van't Land (Eds.), *Higher education's response to the covid-19 pandemic building a more sustainable and democratic future* (pp. 137–146). Strasbourg, France: Council of Europe Publishing.

Milasinovic, S. and Kesetovic, Z. (2008). Crisis and crisis management: A contribution to a conceptual & terminological delimitation. *Megatrend Review, 5*(1), 167–186.

Morley, L. (2003). *Quality and power in higher education.* Maidenhead: Open University Press.

Navarro-Espinosa, J. A., Vaquero-Abellán, M., Perea-Moreno, A.-J., Pedrós-Pérez, G., Aparicio-Martínez, P., and Martínez-Jiménez, M. P. (2021). The higher education sustainability before and during the covid-19 pandemic: A Spanish and Ecuadorian case. *Sustainability, 13*(6363), 1–22.

OECD. (2020). *The impact of covid-19 on education insights from education at a glance 2020.* OECD Report.

Orlowski, M., Mejia, C., Back, R., and Fridrich, J. (2021). Transition to online culinary and beverage labs: Determining student engagement and satisfaction during covid-19. *Journal of Hospitality & Tourism Education, 33*(3), 163–175.

Özsoy, S. (2011). Bilginin metalaşma süreci: Eğitimdeki yapısal dönüşüme ilişkin bazı çıkarımlar. F. Ercan ve S. Korkusuz Kurt (Ed.), *Metalaşma ve iktidarın baskısındaki üniversite* (ss. 121–142). İstanbul: SAV Yayınları.

Park, M. and Jones, T. (2021). Going virtual: The impact of covid-19 on internships in tourism, events, and hospitality education. *Journal of Hospitality & Tourism Education, 33*(3), 176–193.

Pokhrel, S. and Chhetri, R. (2021). A literature review on impact of covid-19 pandemic on teaching and learning. *Higher Education for the Future, 8*(1) 133–141.

Pusser, B. and Marginson, S. (2013). University rankings in critical perspective. *The Journal of Higher Education, 84*(4), 544–568.

Qiang, Z. (2003). Internationalization of higher education: Towards a conceptual framework. *Policy Futures in Education, 1*(2), 248–270.

Recio, S. G. and Colella, C. (2020). *The world of higher education after covid-19: How covid-19 has affected young universities.* Young European Reserach Universities Report. Brussels: YERUN. Retrieved from https://www.yerun.eu/wp-content/uploads/2020/07/YERUN-Covid-VFinal-OnlineSpread.pdf (Access Date: December 7, 2021).

Reimer, D. and Jacob, M. (2010). Differentiation in higher education and its consequences for social inequality: Introduction to a special issue. *Higher Education, 61*(3), 223–227.

Salmi, J. (2002). Facing the challenges of the twenty-first century. *Perspectives: Policy and Practice in Higher Education, 6*(1), 8–12.

Salmi, J. (2020). *Covid's lessons for global higher education.* Indianapolis: Lumina Foundation.

Seo, S. and Kim, H. J. (2021). How Covid-19 influences hospitality and tourism education: Challenges, opportunities, and new directions. *Journal of Hospitality & Tourism Education, 33*(3), 147–147.

Sheldon, P., Fesenmaier, D., Woeber, K., Cooper, C., and Antonioli, M. (2008). Tourism education futures, 2010–2030: Building the capacity to lead. *Journal of Teaching in Travel & Tourism, 7*(3), 61–68.

Slaughter, S. and Laslie, L. (1997). *Academic capitalism.* Baltimore-London: The John Hopkins University Press.

Soroya, S. H., Mohsin, A. R., Zuhair, A., Farhan, M., Khalid, M., and Muhammad A. (2020). Emergency management in higher education during COVID-19 pandemic: A phenomenology inquiry comparing a developed and developing country. *Library Philosophy and Practice (e-journal), 4720,* 1–30.

THE (2020). *University industry collaboration the vital role of tech companies' support for higher education research.* Times Higher Education Consultancy Report, 1–22. Retrieved from https://www.timeshighereducation.com/sites/default/files/the_consultancy_university_industry_collaboration_final_report_051120.pdf (Access Date: December 7, 2021).

Timur, T. (2000). *Toplumsal değişme ve üniversiteler.* Ankara: İmge Kitabevi.

Tiwari, P., Séraphin, H., and Chowdhary, N. (2020). Impacts of COVID-19 on tourism education: Analysis and perspectives. *Journal of Teaching in Travel & Tourism, 0,* 1–26.

Todorova, N. and Bjorn-Andersen, N. (2011). University learning in times of crisis: The role of IT. *Accounting Education, 20*(6), 597–599.

Torres, C. A. and Rhoades, R. A. (2006). Introduction: Globalization and higher education in the Americas. In R. A. Rhoades and C. A. Torres (Eds.), *The* university, state, and market: The political economy of globalization in the Americas (pp. 3–38). Stanford, CA: Stanford University Press.

Tribe, J. (2001). Research paradigms and the tourism curriculum. *Journal of Travel Research, 39*(4), 442–448.

Üsdiken, B., Divarcı Çakmaklı A., and Topaler, B. (2017). Devlet ve vakıf üniversitelerinde "Strateji" 1982-2014. *Yönetim Araştırmaları Dergisi, 13*(1-2), 8–40.

Wissema, J. G. (2009). *Towards the third generation university: Managing the university in transition.* Northampton: Edward Elgar Publishing.

Zaglul, J., Sherrard, D., and Juma, C. (2006). Higher education in economic transformation. *International Journal of Technology and Globalisation, 2*(3/4), 241–251.

Zhong, Y. Y. S., Busser, J., Shapoval, V., and Murphy, K. (2021). Hospitality and tourism student engagement and hope during the covid-19 pandemic. *Journal of Hospitality & Tourism Education, 33*(3), 194–206.

List of Figures

Chapter 3
Fig. 1: Worldwide Covid-19 Monthly Cases (Total and New) (February-December 2020) 67
Fig. 2: International Tourist Arrivals by Month 2020 (Change %) 68

Chapter 4
Fig. 1: Arrivals (Million People) and Receipts ($ Billion) in Post-crisis Years for Thailand 89
Fig. 2: Thailand's Progress in World Rankings 89
Fig. 3: World Ranking of Tourism Expenditures and Receipts of Greece in 2008–2017 95

Chapter 5
Fig. 1: Scale Efficiency 106

Chapter 6
Fig. 1: Timeline for Event Study 128

Chapter 7
Fig. 1: SWOT/SWOC Analysis for Online Learning in Times of Crisis 153
Fig. 2: Emergency Model in Higher Education 157

Chapter 2
Graph. 1: Biggest Perceived Risk Threats to Global Society 33
Graph. 2: Number of Terrorist Attacks Worldwide 2006–2019 40
Graph. 3: Regional Breakdown of Travel Restriction 48
Graph. 4: Opinion on Vaccination among Business Travel Professionals 50
Graph. 5: International Tourist Arrivals 2019-2020-2021 Jan-Mar 51
Graph. 6: Worldwide Revenue with Passengers in Air Traffic from 2005 to 2021 52

List of Tables

Chapter 2
Tab. 1: The Travel and Tourism Competiveness Report 2019 Safety and Security 30
Tab. 2: Chronology of Terrorist Attacks against Tourists 43
Tab. 3: International Tourist Arrivals Between 2000 and 2001 46

Chapter 3
Tab. 1: Pandemic Support for Tourism Sector in Selected Countries* (2020) 70
Tab. 2: Economic Indicators for Tourism in Turkey (2011–2020) 71
Tab. 3: Domestic Travel Indicators in Turkey (2011–2020) 74
Tab. 4: Strengths and Weaknesses, Opportunities and Threats of Domestic Tourism in Turkey before and after Covid-19 during the Pandemic Process 76

Chapter 4
Tab. 1: International Tourism Receipts ($ Billion) 85
Tab. 2: Russia's International Tourism Receipts and Expenditures ($ Billion) 90
Tab. 3: Tourism Receipts and Expenditures of some EU Countries for the Years 2008–2019 94

Chapter 5
Tab. 1: Definition of the Variables 108
Tab. 2: Descriptive Statistics 109
Tab. 3: Efficiency Scores of European Union Countries for Three-Year Periods 110
Tab. 4: Ranking Result of EU Countries 112
Tab. 5: Reference Countries and λ Density Values of EU Countries for 2008–2010 Period 113
Tab. 6: Actual and Target Input Values of Inefficient EU Countries for 2008–2010 Period 114
Tab. 7: Actual and Target Input Values of Inefficient EU Countries for 2011–2013 Period 116
Tab. 8: Actual and Target Input Values of Inefficient EU Countries for 2014–2016 Period 117
Tab. 9: Actual and Target Input Values of Inefficient EU Countries for 2017–2019 Period 118

Chapter 6

Tab. 1:	The List of Companies in City Indices	127
Tab. 2:	Cumulative Average Abnormal Returns for Event Window (−1; 1)	131
Tab. 3:	Cumulative Average Abnormal Returns for Event Window (−2; 2)	133
Tab. 4:	Cumulative Average Abnormal Returns for Event Window (−5; 5)	134
Tab. 5:	Cumulative Average Abnormal Returns for Event Window (−10; 10)	135
Tab. 6:	Cumulative Average Abnormal Returns for Event Window (−20; 20)	136
Tab. 7:	Cumulative Average Abnormal Returns for Event Window (−1; 1)	139
Tab. 8:	Cumulative Average Abnormal Returns for Event Window (−2; 2)	140
Tab. 9:	Tab. 3: Cumulative Average Abnormal Returns for Event Window (−5; 5)	141
Tab. 10:	Cumulative Average Abnormal Returns for Event Window (−10; 10)	142
Tab. 11:	Cumulative Average Abnormal Returns for Event Window (−20; 20)	143

Chapter 7

Tab. 1:	Conflicts/Contradictions in the Change Paradigms of Higher Education	150

Notes on Contributors

Alper ATEŞ was born in Elazığ/Turkey in 1979. He completed his primary, secondary and high school education in Konya. He received his bachelor's, degree in management from Istanbul University and master's doctoral degree in management education from Selçuk University. Working in the fields of Tourism, Tourism Marketing, Tourism Management, Tourism and Environment and Tourism and Technology, Associate Professor Alper ATEŞ has published many articles, papers, books and book chapters in these fields and is the vice editor of Çatalhöyük - International Journal of Tourism and Social Research (Turkey). He is still working at the Selçuk University Faculty of Tourism.

Ali AVAN is the vice dean of Faculty of Tourism at Afyon Kocatepe University. He received his bachelor's degree in hospitality and tourism management at Mersin University, his master's degree in tourism management from Afyon Kocatepe University and his Ph.D. in business administration from Afyon Kocatepe University. He is associate professor in Tourism Management department now, and his areas of research include consumer behavior in tourism, services marketing, tourism marketing and sustainability in tourism.

Ahmet BAYTOK is the head of department at the Tourism Management department at Afyon Kocatepe University and has received his associate professorship in Tourism. He currently interests in the research of leadership in hospitality, organizational behavior, management and sustainability in tourism.

Bayram Şahin was born in 1978. In 2002, he graduated from Balıkesir University School of Tourism and Hotel Management, Travel Management Department. He received his master's degree in 2004 and his doctorate degree in 2011 from Balıkesir University, Institute of Social Sciences, Tourism Management and Hotel Management Department. The author, who received the title of associate professor in the field of "Tourism Management" in 2017, has experience such as coordinator and project manager in various EU projects as well as many scientific studies, book and journal editors. His main areas of study are tourism management, travel management, tourism ethics. He currently works as a faculty member at Balıkesir University Faculty of Tourism and as an assistant manager at Balıkesir University Institute of Social Sciences. Şahin is married and has two children.

Ibrahim Tolga ÇOŞKUN was born in Isparta in 1986. He completed his primary, secondary and high school education in Isparta. He graduated from Cumhuriyet University Education Faculty, as a high school math teacher in 2012 with honors. Between 2012 and 2014, he worked as a dormitory management officer at KYK, in Dinar/Afyon. In 2014, he started to work as a research assistant at Çukurova University, Department of Business Administration. He completed his Master of Business Administration at the same university. He is still doing his PhD. During his PhD education, he studied and worked at the University of Flensburg, Germany. He also holds a bachelor's degree from Anadolu University Management Information Systems. COSKUN, who has thesis and articles on Optimization, Multi-Criteria Decision Making Techniques, Scheduling and different operations research issues, has adopted an interdisciplinary approach in its studies. He is married and has one child.

Selda GÜVEN was born in 1982 in Turkey. She completed her primary and secondary education in Ankara and her high school education in Çanakkale. She graduated from Akdeniz University Hospitality Management Department in 2004. The author, who completed his master's degree at Çanakkale Onsekiz Mart University and worked in the accommodation industry for many years, is still doing his doctorate in Tourism Management at Balıkesir University. Since 2017, she has been working as a lecturer in Çanakkale Onsekiz Mart University Ezine Vocational School, Travel Tourism and Entertainment Services Department. She is doing research in the field of tourism.

Berna KIRAN BULĞURCU has been working as an associate professor at Çukurova University, the Faculty of Economics and Administrative Sciences since 2016. She received her bachelor's and master's degree in business administration from Çukurova University. During her graduate education years, she got education from Sweden Linköping University and Istanbul Technical University at different times. In 2014, she completed her Ph.D. in operations research at Çukurova University. Assistant Prof. Berna Kıran Bulğurcu, who has studied about fuzzy logic, neural network and multi-criteria decision-making techniques, attaches importance to interdisciplinary approach in her studies.

Ömür Hakan KUZU was born in Konya/Turkey in 1976. He received a BSc degree in Public Administration from Faculty of Political Sciences (Mülkiye), Ankara University and the MSc & PhD degrees in Business Administration from Selçuk University. Currently, he is working as an Associate Professor at Selçuk University, Beyşehir Ali Akkanat Faculty of Tourism. His research interest areas include higher education (HE) management, policy, and system. He has published some articles and chapters on topics such as managerialism and digitalization in HE, strategic management in HE, quality and internationalization in HE, world-class universities, European Higher Education Area (Bologna Process), and tourism higher education issues.

İbrahim MISIR completed his undergraduate education at Akdeniz University, Alanya Faculty of Business Administration, Department of Tourism Management in 2013. He worked in various hotels and travel agencies in the Antalya region in between 2010 and 2014. In 2016, he started to work as a research assistant in the Department of Tourism and Hotel Management at Osmaniye Korkut Ata University Kadirli School of Applied Sciences. For his postgraduate education, he was assigned to Balıkesir University Institute of Social Sciences, Department of Tourism Management. Misir, who continue his education and duty in the same department, is doing research in the field of tourism.

Nesrin ÖZKAN was born in Bursa, Turkey. The researcher received the bachelor degree in Economics in English from Anadolu University; master's degree is in Business Administration from Bursa Uludağ University, and the doctoral degree is in Accounting and Finance from Bursa Uludağ University. Dr. Özkan has many publishing in journals those are indexed in SSCI and ESCI about the impact of Covid-19 on financial markets, market anomalies, investment strategies, multifactor asset pricing models, economic policy uncertainty and crypto currency market. Professor's forthcoming publishings will be about financial stress and risk appetite in markets. She is currently the faculty member of İstanbul Atlas University.

Elbeyi PELİT was born in Gümüşhane/Turkey in 1978. He completed his primary, secondary and high school education in Gümüşhane. He received his bachelor's, master's degree in tourism management from Sakarya University and doctoral degree in tourism management education from Gazi University. Working in the fields of Tourism, Tourism Education, Tourism Management, Human Resources Management in Tourism and Tourism policy, Professor Elbeyi PELİT has published many articles, papers, books and book chapters in these fields and is the editor of the *Journal of Tourist Guide*, *Journal of Contemporary Tourism Research* and *Afyon Kocatepe University Journal of Social Sciences* (Turkey). He is still working at the Afyon Kocatepe University Faculty of Tourism.

H. Hüseyin SOYBALI was born in Afyonkarahisar province of Turkey in 1967. He completed his primary, secondary and high school education in Afyonkarahisar. He received his bachelor's degree from Uludag University, Balikesir School of Tourism and Hotel Management. He received his MSc degree in Tourism Management from the University of Surrey and PhD degree from Bournemouth University in England, UK. His main research interest areas include tourism development, tourism management, tourism policy and planning, human resources management and tourism education. Professor SOYBALI has published a book, many book chapters, articles and papers. He is working at the Afyon Kocatepe University, Faculty of Tourism as a senior professor and vice dean.

Selcen Seda TURKSOY was born in İzmir/Turkey in 1978. She graduated from English Language Education Department, Dokuz Eylul University, Turkey in 2000. She obtained her master's degree and doctoral degree in tourism management from the same university, respectively in 2005 and 2017. She is currently working at Ege University Cesme Faculty of Tourism as an Assistant Professor. She has published articles, papers, and book chapters in the field of tourism. Her research interests are human resources management and organizational behaviour in tourism as well as sustainable tourism.

Özcan ZORLU is the co-head of Tourism Guidance department at faculty of Tourism at Afyon Kocatepe University. He received his bachelor's degree in hospitality and tourism management at Balıkesir University and his master's degree in tourism management from Balıkesir University. He has received his associate professorship in tourism and written widely on specifically organizational behavior, knowledge management and alternative tourism activities in tourism.

About Editors

Elbeyi PELİT is Prof. Dr. at Faculty of Tourism, Afyon Kocatepe University, Afyonkarahisar (Turkey)

Hasan Hüseyin SOYBALI is Prof. Dr. at Faculty of Tourism, Afyon Kocatepe University, Afyonkarahisar (Turkey)

Ali AVAN is Assoc. Prof. at Faculty of Tourism, Afyon Kocatepe University, Afyonkarahisar (Turkey)

www.ingramcontent.com/pod-product-compliance
Ingram Content Group UK Ltd.
Pitfield, Milton Keynes, MK11 3LW, UK
UKHW021842210426
5322IPUK00022B/409